Staying Italian

Dear Michael,
Many, many thanks
for your role in
this.
– Jordan

Also in the series:

New York Undercover: Private Surveillance in the Progressive Era *by Jennifer Fronc*

African American Urban History since World War II *edited by Kenneth L. Kusmer and Joe W. Trotter*

Blueprint for Disaster: The Unraveling of Public Housing in Chicago *by D. Bradford Hunt*

Alien Neighbors, Foreign Friends: Asian Americans, Housing, and the Transformation of Urban California *by Charlotte Brooks*

The Problem of Jobs: Liberalism, Race, and Deindustrialization in Philadelphia *by Guian A. McKee*

Chicago Made: Factory Networks in the Industrial Metropolis *by Robert Lewis*

The Flash Press: Sporting Male Weeklies in 1840s New York *by Patricia Cline Cohen, Timothy J. Gilfoyle, and Helen Lefkowitz Horowitz in association with the American Antiquarian Society*

Slumming: Sexual and Racial Encounters in American Nightlife, 1885–1940 *by Chad Heap*

Colored Property: State Policy and White Racial Politics in Suburban America *by David M. P. Freund*

Selling the Race: Culture, Community, and Black Chicago, 1940–1955 *by Adam Green*

The New Suburban History *edited by Kevin M. Kruse and Thomas J. Sugrue*

Millennium Park: Creating a Chicago Landmark *by Timothy J. Gilfoyle*

Additional series titles follow index

Staying Italian: Urban Change and Ethnic Life in Postwar Toronto and Philadelphia

Jordan Stanger-Ross

The University of Chicago Press :: Chicago and London

Jordan Stanger-Ross is assistant professor of history at the University of Victoria, British Columbia.

The University of Chicago Press, Chicago 60637
The University of Chicago Press, Ltd., London

Printed in the United States of America

18 17 16 15 14 13 12 11 10 09 1 2 3 4 5

ISBN-13: 978-0-226-77074-1 (cloth)
ISBN-10: 0-226-77074-5 (cloth)

Library of Congress Cataloging-in-Publication Data

Stanger-Ross, Jordan.
Staying Italian : urban change and ethnic life in postwar Toronto and Philadelphia / Jordan Stanger-Ross.
p. cm.
Includes bibliographical references and index.
ISBN-13: 978-0-226-77074-1 (cloth : alk. paper)
ISBN-10: 0-226-77074-5 (cloth : alk. paper)
1. Italian Americans—Pennsylvania—Philadelphia—Ethnic identity—History—20th century. 2. Italians—Ontario—Toronto—Ethnic identity—History—20th century. 3. Ethnicity—United States—History—20th century. 4. Ethnicity—Canada—History—20th century. 5. Ethnic neighborhoods—United States—History—20th century—Case studies. 6. Ethnic neighborhoods—Canada—History—20th century—Case studies. I. Title.
E184.I8S73 2010
305.895'1073074811—dc22

2009018815

♾ The paper used in this publication meets the minimum requirements of the American National Standard for Information Sciences—Permanence of Paper for Printed Library Materials, ANSI Z39.48-1992.

Contents

Illustrations

Figures

Maps

Acknowledgments

The completion of this project has brought many satisfactions, not the least of which is a fuller appreciation of the depth of gratitude expressed in a writer's acknowledgments. This book began as a dissertation at the University of Pennsylvania. My supervisor, Michael Katz, provided unfailingly precise, timely, and judicious advice throughout my graduate career and beyond. His ability to ask nagging questions, questions that continued to haunt my thoughts long after I first imagined them answered, vastly improved every facet of this book. In addition, Thomas Sugrue, Michael Zuckerman, Ewa Morawska, and Daniel Richter read and commented on the entire dissertation. Their comments invariably lifted this project along when it threatened to flag. The quantitative aspects of this work benefitted from the help of Mark Stern and Amy Hillier, both of whom were tremendously responsive and patient in the face of my considerable demands. Additional technical assistance was provided by Lauris Olson and Laurie Allen at the University of Pennsylvania, and Laine G. M. Ruus, Sushil Kumar, and Marcel Fortin at the University of Toronto.

At Penn, I found myself surrounded by a remarkable group of graduate students. Julia Rabig's creative, eclectic intelligence remains an invaluable resource for my work, just as her delightful company has proved indispensable at conferences. Thanks also to Benjamin Field, Leah Gordon, and Domenic Vitiello, each of whom read parts of my dissertation or closely related work. Conversations with others in the department also sharpened my thinking. Special thanks to Francesca Bregoli, Christina Collins, Amy Garrett, Christopher Klemek, Ben Mercer, Richard Ninnis, Anne Oravetz, Katie Paugh, James Saporito, Peter Siskind, Lorrin Thomas, and Francis Ryan.

Since my return to Canada, generous institutions and welcoming colleagues have helped to nurture this book. When I was still a student, Rick Halpern invited me to spend two years at the Centre for the Study of the United States at the University of Toronto. While there, I benefited from Steve Penfold's cheerful companionship at street processions, calming advice about the job market, and frequent reminders that the book ought to include quotes as well as numbers. Franca Iacovetta provided advice early in this project and, toward the end, miraculously excavated tables from her own dissertation for my use. As supervisor of my postdoctoral fellowship, John Zucchi read the entire manuscript, providing useful advice and much appreciated support.

At the University of Victoria, the flexibility of the dean of humanities, Andrew Rippin, as well as department chairs Eric Sager and Tom Saunders, afforded me extraordinary research and writing opportunities during the first three years of my appointment. My colleagues in the history department have extended a warm welcome. Special thanks are owed to Peter Baskerville, Sara Beam, Greg Blue, John Lutz, Lynne Marks, Eric Sager, Elizabeth Vibert, and Wendy Wickwire, all of whom have read my work, hosted my family, or otherwise made me feel especially welcome at the university and in Victoria. On very short notice, Jason Colby read the entire book manuscript in its final stages and has graciously accepted responsibility for any shortcomings that still remain. Many other scholars were also generous with their time and responsive to this work. Richard Harris, Chad Gaffield, Russell Kazal, and Thomas Guglielmo read the entire manuscript and provided important feedback. Conversations and exchange with Christopher Friedrichs, Donna Gabaccia, Steven High, Robert Lewis, Robert McDonald, Bruno Ramirez, Coll Thrush, and Henry Yu improved my work while also making it more enjoyable.

This work also benefited from the help of archivists at the Philadelphia City Archives, the Philadelphia Archdiocesan Historical Research Center, the City of Toronto Archives, and the Archives of Ontario. The Archives of the Roman Catholic Archdiocese of Toronto deserves particular mention. Under the directorship of Marc Lerman, the archive is a model of efficiency and professionalism. Fathers Gregory Botte, Herb Sperger, Arthur I. Taraborelli, Anthony J. D'Angelico, and Gary Pacitti trusted me enough to launch this project. Residents of the Italian parishes in both cities were generous with their time and open in their recollections. The hours that I spent reminiscing with longtime Italian residents of both cities were the highlights of this process. Vince Pietropaolo deserves special thanks for a lengthy and informative interview and for his wonderful photographs. Thanks also to Francesca Schembri for sharing considerable insight and a memorable dinner on a cold Toronto evening. Pasang Thachhoe at the Multicultural Historical Society of Ontario provided gracious assistance during the many hours I spent in his company listening to the recollections of Italian Torontonians.

Various agencies have supported this research during the past decade. Thanks are owed to the University of Pennsylvania's Graduate Fellowships, Benjamin Franklin Fellowships, and Annenberg Graduate Fellowships; the Social Science and Humanities Research Council of Canada; internal grants at the University of Victoria; as well as to the Catholic Historical Association, the Salvatori Research Award at the University of Pennsylvania, and the Association for Canadian Studies in the United States.

My experience with the University of Chicago Press has exemplified the best of academic publishing. Tough but supportive anonymous readers vastly improved this book. Harry Johnson worked quickly and cheerfully to make my maps more attractive and communicative. Timothy Gilfoyle provided careful and frank editing, and Robert Devens gently pushed the project forward from the start. I am simply delighted to publish the book in a series that I have long admired.

There are also those whose support defies cataloging. My closest friends and family have been deeply enmeshed in this work, suffering my attempts to reinvent household modes of production, entertaining my efforts to clarify thoughts by speaking them aloud, and providing frequent and abundant reason for celebration and distraction. Jeff Allred, Gretchen Aguiar, Eric Adams, and Sarah Krotz have accompanied me throughout this process. I am profoundly grateful for our friendships. My in-laws, and in particular Sheila and Melvin Stanger, have welcomed me and my work with warmth and a sometimes daunting faith in my abilities. Living near my family in Toronto for a period of this work was a great pleasure. My siblings, Corey and Ilana, and their spouses, Diana and Dan, along with my wonderful niece Maia, tremendously enriched the choreography of my daily life for the period that

I spent among them. My grandmother, Edith Ross, has shown unfailing interest in this project, even if she sometimes thought that my time ought to be spent in front of a classroom rather than a computer. My parents, Hildy and Michael Ross, have continued to be models of both scholarship and family. My mother was always on hand for emergency statistical analyses, and their seaside home in Nova Scotia was the site of my best and most enjoyable work. Ilana Stanger-Ross has filled the last decade and a half of my life with love, inspiration, beauty, and laughter. She gave this manuscript its final read, even as she balanced the demands of her own forthcoming book and her studies to become a midwife. I am lucky to share my life with someone who does so much so well. It is because of her, and now our delightful daughters Tillie and Eva, that I write these sentences brimming with happiness.

Introduction

Just as I was beginning to grasp the full direction of this project, Sister Carmel Marie of Annunciation Parish in South Philadelphia surprised me at what I thought would be an individual interview by inviting along four other women. A formal interview of the old friends, who ranged in age from their fifties to their nineties, was clearly out of the question, so I pressed record, sat back, and listened. As the women reminisced about growing up in the city's oldest Italian neighborhood, they sketched the social geographic networks that would become the core of this book.

Recalling their youthful romances during World War II and the decades that followed, the women described powerful bonds among local residents. One gestured with her hands, "I lived on this corner, and if you walked to the next corner was my husband, and we got to know each other and we went out." Another explained: "[Y]ou all knew their mothers and fathers and aunts and uncles, you didn't have to be introduced." A third met her husband in her family store: "His mother used to come in and buy the groceries, and he would come in once in awhile . . . we started going to skating parties and the dances and all, and that was that. I was only fifteen, he was fourteen . . . and I'm still dancing."[1]

Audrey Geniole was born in Italy but lived most of her

life in Toronto; she told a different story of courtship when A. McPeek interviewed her in 1977. Geniole described her experiences in Toronto's Little Italy in terms that often echoed those of the South Philadelphians: "Everybody knew everybody, everybody visited each other and on a Saturday night it was like a street dance."[2] Yet a circuitous path had led her to marriage. Geniole met her husband-to-be in the 1940s. Although she was then living outside of Little Italy with her older brother, she recalled, "[I] used to come down [to Little Italy] a lot because I had a girlfriend . . . we used to hang out around the corner." It was during one of these visits to Little Italy that a young man pulled up on a motorcycle, "and of course we went for a horse ride on the motorcycle and the next thing I knew I was going with my husband. Mind you, he didn't like the idea of going steady, but we made it."[3] Geniole's story involved movement in and out of Little Italy. Initially an immigrant to the area, she met an Italian man when she came back to Little Italy from outside, and her husband was passing through the neighborhood when they met.

As the women of Italian South Philadelphia and Toronto's Little Italy described their encounters with their husbands, they illuminated more than individual romances.[4] Each of the women took an ethnic Italian as her spouse. Their stories typify the gatherings of Italian ethnics within the two neighborhoods. The purposes for gathering varied. Italians came together at the dance halls and street corners where they met their spouses, at the church pews and religious processions where they offered prayers, and at the banks, stores, and offices where they worked and conducted business. Gathering with coethnics in public and private places, Italians in both neighborhoods created community.

Urban environments shape daily as well as life-changing choices in ways that often go unnoticed by city dwellers. As they moved within the cities they inhabited, the Italians of Toronto and Philadelphia built different kinds of communities. They gathered in different ways. The Italians of Annunciation Parish in South Philadelphia connected with fellow ethnics through stable local ties, whereas the connections among residents of Toronto's Little Italy were characterized by relocation and neighborhood flux. These differences might not have been noticed by the residents themselves; the small decisions of daily life—where precisely one gathers for a social or religious occasion and with whom—can seem of little consequence. However, the large and small choices that shaped social lives within the two Italian neighborhoods bore the imprint of vastly different postwar cities.

The two Italian neighborhoods offer intriguing sites for the consideration of urban community. At the close of World War II, Italian South Philadelphia and Toronto's Little Italy shared much in common. Both neighborhoods had grown out of the migration of millions of Italians to North American industrial cities in the late nineteenth and early twentieth centuries. Both have been described by previous historians as sites of local, neighborhood ties—places where common languages, needs, and interests brought Italians together in their social lives, economic endeavors, and cultural expressions. The gatherings of Italian ethnics at local Roman Catholic churches, cafes, bars, and grocery stores fostered the robust street life and dense local ties that characterized urban Italian ethnicity in North America.[5] At first glance, the recollections of women who lived in the two neighborhoods suggest that Italian urban life continued in this fashion well after World War II. However, closer scrutiny reveals that the personal lives of Italians reflected the changing neighborhoods and cities that they inhabited.

Philadelphians confronted the hardships of many cities in the United States. Policy decisions at all levels of government, widespread prejudice within labor and housing markets, and economic restructuring all contributed to the racial and economic segregation of postwar cities.[6] Division prevailed in the country's declining industrial centers as well as within cities that saw growing concentrations of capital and workers.[7] Philadelphia, a city that lost jobs and population while splintering along racial and class lines, fit firmly within this national trend. Italian South Philadelphians built their social lives within the setting that two prominent sociologists have provocatively described as an "American Apartheid."[8]

The place of older ethnic neighborhoods within postwar urban restructuring and racial division in the United States is only dimly understood. A consensus of historians and sociologists has emphasized the diminishing salience of distinctions among Americans of European origins in the middle decades of the twentieth century. However, evidence from postwar cities suggests that ethnicity, and perhaps in particular Italian ethnicity, retained considerable importance. By 1980, when many Italian Americans had intermarried and moved to ethnically diverse neighborhoods, some 570,000 of their coethnics continued to live in pockets where the majority of the population reported Italian ethnic origins. More than 260,000 of these lived in New York, especially in Brooklyn. In addition, 55,000 Italians in Philadelphia, 50,000 in Providence, Rhode Island, and 40,000 in Boston comprised majorities in their local surroundings. These persistent Italian neighborhoods—along with smaller enclaves in Chicago, New Haven, New Jersey (in particular where it bordered New York and Philadelphia), and elsewhere—complicate the scholarly view that Italian ethnicity became a symbolic or individual identity in the postwar era. In many traditional areas of immigrant settlement, ethnicity continued to determine where, and among whom, Italian Americans lived.[9]

This study examines the experience of ethnicity in one such neighborhood, comparing social life in Italian South Philadelphia with a very different site of Italian settlement, Toronto's Little Italy. Attention to social experience—residential patterns and real estate transactions, religious and associational practices, marriage choices, and labor force participation—reveals the changing, but still powerful, role of ethnicity in urban neighborhoods. In Canada, where mass immigration continued throughout the postwar era, the importance of ethnicity to city history is more readily observed. But in the United States too—where postwar scholarship has focused on the politics of race rather than the social history of ethnicity—Italian Americans, organized as such, continued to dominate important urban pockets.

In emphasizing the importance of ethnicity, I do not seek to diminish the role of race in postwar America. Instead, comparison of Toronto and Philadelphia confirms the importance of urban conflict in the United States. Daily life in South Philadelphia bore the imprint of the city's racial and economic divide. However, rather than displacing Italian identities and affiliations, racial tension reinforced ethnic bonds. Italian South Philadelphians used ethnicity to navigate urban crisis.[10]

As this study intervenes in discussions of ethnicity and race in the postwar United States, it simultaneously contributes to related scholarship on urban religion. Comparison of Toronto and Philadelphia cautions against the assumption, which has recently crept into the literature on postwar North American cities, that Catholicism inherently encourages defensive neighborhood attachments. This book does not dispute the finding that Catholics in postwar cities in the United States, and in particular the urban North, often defended urban turf. On the contrary, Italian South Philadelphians displayed the defensive localism evident in other postwar Catholic neighborhoods.[11] However, comparison with Toronto confirms, as Robert Orsi has argued, that religion is "fundamentally and always *in* history."[12] The patterns of Italian life in Toronto demonstrate that defensive localism was only one way that Catholicism could shape twentieth-century city life. In Toronto, Italian Catholics used ethnicity and religion differently as they negotiated different urban challenges. Italian ethnics defending turf in Philadelphia participated in the particular conflicts of the postwar urban United States; in Toronto Italian ethnicity and Catholicism operated in a very different fashion.

By contrast to Philadelphia, postwar Toronto offered a prosperous setting for Italian social experience. The decades after World War II brought dramatic economic and demographic growth in southern Ontario, partially as a result of significant postwar immigration.[13] Toronto made a rapid and successful transition from an industrial to a service economy, emerging in the last decades of the twentieth century as the "command center" of Canadian financial and corporate activities.[14] The metropolitan government, a regional umbrella encompassing Toronto and its surrounding municipalities, linked the fate of the center city with the wider region. Economic growth in some two hundred square miles surrounding the city funded services and infrastructure at the urban core.[15] Italian Torontonians—many of them immigrants struggling to maximize limited resources—used ethnicity in their attempts to share in citywide prosperity.

With thousands of new immigrants, rising property values, and abundant nearby work opportunities, Toronto's Little Italy exemplified the wider dynamics of the postwar city. And yet, as in South Philadelphia, the history of the area in the postwar era remains only poorly understood. Indeed, historians have yet to explain the most startling development in Little Italy: its rapid decline as a center of Italian settlement. With thousands of Italians arriving annually in Canada in the 1950s and 1960s, the Italian population of Little Italy reached 12,000 in 1971, and two of three among them reported birth origins in Italy.[16] However, the following decade brought a sudden decline in the Italian residential enclave. In contrast to the stability of Italian South Philadelphia, the Italian presence surrounding College Street quickly thinned. The number of residents in Little Italy reporting Italian origins fell by almost 60 percent in the decade after 1971 and continued to drop thereafter. While second- and third-generation Italians in South Philadelphia remained entrenched in their old neighborhood, new immigrants in Toronto were on the move. The area remained an important gathering place for Italian ethnics—most notably, hundreds of thousands continued to congregate annually for the Good Friday procession—but the area ceased to house large numbers of Italians.

Comparative study helps to explain the rapid dissolution of Toronto's Little Italy. This book argues that the history of the area reflects the particular dynamics of Italian ethnicity in postwar Toronto. While South Philadelphians used ethnicity defensively to preserve their neighborhood, Italian Torontonians mobilized ethnic bonds in their attempts to seize the opportunities available in the city. These contrasting case studies reveal ethnicity as a pliable tool; Italians in the two contexts responded to very different urban imperatives.

This book focuses on pivotal choices made by Italian ethnics. Taken together, decisions about housing, religious practices, marriage, and workforce participation, each of which I consider in subsequent chapters, shaped much of social experience. Important life choices also reflect a much wider constellation of social processes. The choice of a marriage partner, for example, grows out of innumerable social gatherings. Even strangers who meet in a glance across a crowded room must first find themselves in the same crowded room. The patterned life choices of Italian ethnics—choices that shaped social gatherings—illuminate their responses to the constraints and opportunities of urban experience.

Choices about housing, religious practices, and marriage followed a common pattern, with ethnicity playing an important but distinctive role in each city. Residents of the neighborhoods exchanged housing, prayed, and married within heavily Italian circles well after World War II. Rather than receding into a symbolic realm, ethnicity continued to organize social life.

But Italian ethnicity was a different kind of bond in the two historical contexts. In South Philadelphia Italian ethnicity was closely tied to territory. Italian ethnics gathered in ways that reinforced the bonds among residents of their local surroundings, evoking Italian ethnicity as a mechanism for preserving urban turf. Italian South Philadelphians hoped that ethnicity might insulate them against urban decline. In Toronto's Little Italy, ethnicity operated with little regard for neighborhood. Italians in Little Italy developed ongoing, systematic, and intimate ties with coethnics dispersed throughout the metropolitan area and beyond. In prospering Toronto, ethnicity helped Italians seeking economic gain and social connection, but it did so in a geographically elastic or expansive fashion. Rather than functioning as an urban boundary, Toronto's Little Italy was a shared ethnic space, a gathering place for Italians far and wide. The paths that Italian ethnics took en route to one another, the "choreography" of social life in each city, made Italian ethnicity a different social practice in the two places.[17]

Workplace decisions followed a distinctive and surprising trajectory. In South Philadelphia, where neighborhood held strong, ethnicity ceased to play a key role in workplace decisions. As industrial employment declined, Italians in Philadelphia abandoned ethnic labor niches. In Toronto, meanwhile, Italians continued to make use of labor niches, even as they broke free of the local neighborhood. In both places, the social geography of workforce participation diverged from the rest of social life. This development represented a dramatic change in the character of Italian experience, which had previously taken shape in the nexus between work and home.

For the most part, my argument focuses on the character of ethnic bonds. The degree of ethnic association—or conversely, the extent of assimilation—in each locale is not a central focus of this study. I argue that ethnicity remained influential in both Toronto and Philadelphia; Italians continued to gather together in important neighborhoods in both cities. Ethnicity continued to shape urban life. However, rather than detailing the quantity of ethnic participation, the chapters that follow emphasize its qualitative characteristics. The different trajectories of the two neighborhoods, I argue, were reflections of the way that ethnicity was practiced in each place rather than the extent to which it was practiced. In postwar South Philadelphia Italian ethnicity guarded neighborhood, whereas in Toronto's Little Italy it operated on a metropolitan scale.

This study contributes to scholarship that has increasingly demonstrated the shifting, dynamic, and contested character of ethnicity. Recognizing that ethnicity is a social practice rather than an immutable attribute, historians have detailed its change over time.[18] This book concentrates on the differences between localities, a somewhat less well understood facet of the historical contingencies that shape social life. Comparing Italian life in Toronto and Philadelphia, I argue that ethnicity varied significantly by place. The social practices that constituted ethnic community varied by North American locality, just as they did by origin groups and over time.

North American cities were transformed in the decades after World War II as suburbanization, segregation, and economic restructuring reshaped the physical and social characteristics of urban life. As these events transpired differently in different cities, the histories of North American urban areas diverged. The new economy and geography of North American life brought prosperity to some cities and steep decline to others.[19] Yet we have little knowledge of how these transformations recast the bonds among the ethnic groups that arrived in North America during the industrial heyday. What ramifications did postwar urban change carry for social connections founded in density and proximity? How did patterns of daily life shift as a result of the changing structure and politics of cities? Comparing Toronto and Philadelphia from the early 1950s to the late 1980s, this study examines two cities as they were reshaped in the postwar era. Differential postwar changes encouraged different forms of ethnic association. The gatherings of Italian ethnics in both cities expressed the interwoven effects of politics, economy, and race in the making of ethnic life.

1 Cities Apart: Toronto and Philadelphia after World War II

Toronto and Philadelphia charted opposite paths during the second half of the twentieth century, setting dramatically different stages for Italian ethnic experience. The two cities reflected the unevenness of postwar urban restructuring in North America, a process that involved the relocation of resources and people as well as the emergence of services as the continent's leading economic sector. In different regions and metropolitan areas these changes took different forms; localities competed with one another to retain residents and attract investment, and the losers struggled while the winners thrived.[1] Toronto stood among the winners in urban restructuring; Philadelphia, at least until the 1980s, stood among the losers.[2]

For the most part, this book focuses on two small neighborhoods, but the wider postwar history of both cities is crucial to the story that I will tell. Residents of Toronto's Little Italy and Italian South Philadelphia responded to developments at citywide, regional, and national levels. Even their most intimate and personal choices bore the imprints of events that lay largely outside of local control. This chapter traces some of the most salient aspects of that wider context, detailing the political, economic, and social dynamics that made Toronto and

Philadelphia dramatically different places in the decades after World War II.

: : :

Housing values in Toronto and Philadelphia epitomized the divergent trends afoot. Despite the slight decline in the city's population, home values in Toronto rapidly rose throughout the postwar era. Indeed, homeowners in the Ontario capital saw prices rise by 50 percent, on average, in every decade. Further, the city of Toronto retained particular value by comparison with its wider metropolitan area. Throughout the postwar period, houses in Toronto's oldest urban neighborhoods were more valuable, on average, than those in newly built suburbs.[3] Philadelphians owned homes in a notably different market. Between 1950 and 1970, the real value of housing in Philadelphia declined as the city suffered through the worst of its postwar struggles (fig. 1). For two decades, postwar Philadelphians saw little, if any, economic benefit from homeownership. Following this protracted stagnation, housing values increased modestly in the 1970s. Then, in the 1980s, as Philadelphia began to emerge from its postwar malaise, real estate values improved significantly. Even after the 1980s, however, Philadelphia values still languished well behind those in Toronto. Further, Philadelphia was disadvantaged within its wider region. In 1980, houses in the surrounding counties were some two to three times more valuable than those in Philadelphia.[4] Keenly felt by residents of neighborhoods such as Italian South Philadelphia and Toronto's Little Italy, housing markets provide telling barometers of the postwar history of both cities.

Urban restructuring unfolded differently in Toronto and Philadelphia in part because of differences between postwar immigration policies in Canada and the United States. Until the mid-1960s, Canada and the United States preserved national quotas and permitted the exclusion of immigrants on the basis of nationality, political persuasion, and a range of other personal characteristics. It was not until the mid-1960s that the two countries significantly reformed their respective immigration laws.[5] However, the two policies differed notably in the number of immigrants they permitted relative to the total national populations of each country. Some 3.5 million immigrants were legally admitted to the United States between the passage of two major postwar immigration acts in 1952 and 1965, amounting to 2 percent of the national population at the outset of the period. During roughly the same years—between immigration acts in 1952 and 1967—Canada admitted

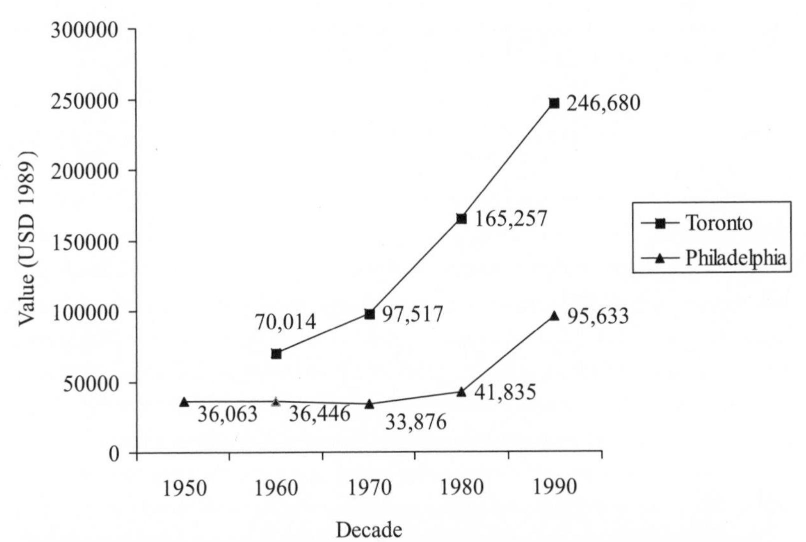

FIGURE 1 Median value of owner-occupied dwellings in Toronto (city) and Philadelphia, 1950–1990. Original values reported as medians, except in the case of the 1981 and 1991 censuses in Canada, which reported average values. SOURCE: U.S. Bureau of the Census, *U.S Census of Housing, 1950* V.I, pt. 5, table 21, 38-52 (Washington, DC: U.S. GPO, 1953); U.S. Bureau of the Census, *U.S. Census of Housing, 1960* V.II, pt. 5, table A1, 137-3 (Washington, DC: U.S. GPO, 1963); U.S. Bureau of the Census, *U.S. Census of Housing, 1970: Metropolitan Housing Characteristics, Philadelphia, PA-NJ Standard Metropolitan Statistical Area,* table A-1, 165-1 (Washington, DC: U.S. GPO, 1972); U.S. Bureau of the Census, *U.S. Census of Housing, 1980: Metropolitan Housing Characteristics, Philadelphia, PA-NJ Standard Metropolitan Statistical Area,* table A-1, 283-1 (Washington, DC: U.S. GPO, 1983); U.S. Bureau of the Census, *U.S. Census of Population and Housing, 1990: Population and Housing Characteristics for Census Tracts and Block Numbering Areas: Philadelphia-Wilmington-Trenton, PA-NJ-DE-MD CMSA, Philadelphia PA-NJ PMSA,* section 1, table 9, 384 (Washington, DC: U.S. GPO, 1993); Dominion Bureau of Statistics, *1961 Census of Canada,* Series CT, "Population and Housing Characteristics by Census Tracts," Toronto, table 2, 26 (Ottawa: Ministry of Trade and Commerce, 1964); Statistics Canada, *1971 Census of Canada,* Census Tract Bulletin, Series B, "Population and Housing Characteristics by Census Tract," Toronto, table 2, 50 (Ottawa: Ministry of Trade and Commerce, 1974); Statistics Canada, *1981 Census of Canada,* Profile Series B, vol. 3, Census Metropolitan Areas with Components, Selected Social and Economic Characteristics, table 1 (Ottawa: Ministry of Trade and Commerce, 1983); Statistics Canada, *1991 Census of Canada,* Profile of Census Divisions and Subdivisions in Ontario, Part B, table 1 (Ottawa: Ministry of Industry, 1994).

2.1 million immigrants, or 15 percent of the national population at the outset of the era.[6]

National differences were exaggerated in Toronto and Philadelphia. Canada's most popular immigrant destination, Toronto shared disproportionately in the flood of newcomers. In the first two decades after World War II, immigrants from Europe, including thousands from Great Britain, Italy, Portugal, Greece, and Germany, poured into the city. By 1971, foreign-born residents accounted for 44 percent of the population in metropolitan Toronto; Italians, with 8 percent of the population, were the largest single immigrant group in the city.[7] As many Torontonians headed to the suburbs, the influx of immigrants to Toronto stabilized the population and kept property values rising. Philadelphia, by contrast, drew few new immigrants from abroad. In 1970, all but 7 percent of the city's population had been born in the United States.[8]

The most conspicuous newcomers in Philadelphia during the immediate postwar decades were not foreign immigrants, but rather black migrants from the American South. Philadelphia's African American population grew by 50 percent in the 1940s alone, partly as a result of natural increase, but also because of the ongoing Great Migration. In 1970, as the peak years of migration came to a close, just over a quarter of the black population in Philadelphia reported birth origins in the southern United States.[9] While this population, totalling some 165,000 people, was small by comparison with the number of immigrants in Toronto, it exceeded the foreign-born population in Philadelphia. As white Philadelphians left the city for the suburbs, black migration into Philadelphia contributed to an increasingly stark racial divide within the metropolitan region. In 1980, African Americans comprised 10 percent or less of the population in most of the counties surrounding the city. In Philadelphia, the black population had risen from 18 percent in 1950 to 42 percent in 1980.[10]

Early postwar patterns persisted after the changes to immigration policy in the 1960s. Toronto remained Canada's dominant immigrant destination, with rising numbers of non-Europeans joining the stream of newcomers to the city.[11] Meanwhile, immigrants remained reluctant to settle in Philadelphia, despite the opening of the United States borders. Even in 1990, after a decade in which more immigrants arrived to the United States than at any time since 1910, Philadelphia's foreign-born population comprised less than 10 percent of the total.[12]

Toronto's appeal to immigrants reflected the city's considerable share in provincial and national prosperity. Cyclical downturns aside, the Ontario economy grew dramatically between the 1950s and 1970s as both labor and capital concentrated in the province. In these early postwar decades the industrial economy held strong. Population growth and women's increased workforce participation brought more hands than ever before to provincial factories. With labor productivity on the rise and capital flowing from abundant foreign and domestic sources, Ontario regularly set new production records in the years prior to 1974 while leading the country in all manner of manufacture.[13] And if the province stood at the center of national prosperity, Toronto was Ontario's industrial core. While the provincial manufacturing workforce grew by an impressive 24 percent in the 1960s, metropolitan Toronto's industrial workers outpaced the rest of Ontario, growing by 35 percent. By 1970, four of ten manufacturing jobs in the province were located in Toronto's metropolitan area.[14] A survey of the city's manufacturers in the mid-1960s found that city workers produced an increasing number of products for sale outside of the local market. Automotive parts, electrical apparatus, industrial machinery, building supplies, food products, and furniture made in Toronto were shipped to every corner of the country as well as across the border to the United States and overseas.[15]

Crucial to the economic health of Toronto and to its prosperity beyond the industrial downturn of the 1970s was a second, simultaneous area of growth—the service sector. While the industrial labor force continued to grow in absolute terms, it relinquished its position of dominance relative to other forms of work. The industrial share of the metropolitan Toronto workforce fell from a high of 35 percent in 1951 to 17 percent in 1991. These proportional losses reflected the dramatic growth of services as much as any weakness of industry.[16] Financial services had established a firm base in Toronto by the mid-1960s, when the metropolitan area claimed 45 percent of the assets in all of the financial institutions in Canada.[17] In the decades that followed, Toronto solidified its position as the "command center" of Canadian financial and corporate activities, becoming the unrivaled hub of Canadian business services and one of the most important financial centers on the continent. By 1991, 42 of 56 international banks in Canada had headquarters in Toronto, as did half of Canada's foreign-owned companies. The nation's top 5 banks, 6 of its top 10 insurance companies, and 193 of the top 500 corporations all placed executive control in Toronto. The city's stock exchange, with more than $67 billion (CAD) traded in 1991, ranked forth in North America and eleventh in the world.[18]

The geography of the service sector advantaged Toronto's downtown. Unlike industrial production, financial services, especially their highest functions and best-paid employees, concentrated in the city center and increased in vitality.[19] By 1985 the service sector accounted for 72 percent of employment in the metropolitan area.[20] Postwar Toronto made a rapid and successful transition to a new economic base.

In Philadelphia, postwar economic history unfolded less fortuitously, beginning with the erosion of the city's industrial labor force. Industrial job losses in Philadelphia averaged 26 percent per decade after 1950, dropping the total manufacturing jobs from over 355,000 to some 100,000 by the late 1980s.[21] The processes that led to job losses in Philadelphia had been underway for much of the twentieth century. An ever increasing number of Philadelphia workers were employed by large, nonlocal firms that ultimately preferred factories outside of the city. Local concentration in nondurable products, such as textiles, tobacco, and food, left Philadelphia especially vulnerable. Such firms, whose smaller operations are more mobile than those of durable goods producers, faced limited costs in relocating to favorable labor, real estate, and market conditions. In the postwar period many of the city's most important industrial employers sought greener pastures in the suburbs, elsewhere in the country, and abroad.[22]

The service economy did much to stem the tide of job losses. Unlike Toronto, however, Philadelphia did not emerge as a major center of the new economy's elite service functions, such as finance and business services.[23] The number of workers in finance, real estate, and insurance increased some 20,000 between 1950 and 1970, but then reached a plateau of just under 70,000 workers. Much more important to the service sector were a range of other occupations—led by jobs in the fields of health care and education—which accounted for almost 200,000 new jobs in the city between 1950 and the late 1980s.[24] By the last decades of the century, growth in these areas imbued Philadelphians with cautious optimism about the economic prospects of their city, a sentiment that had been all but submerged by industrial decline.[25]

While the upturn of the late 1980s will be crucial to the last stages of the history told here, Philadelphia's annual job losses between the mid-1950s and the mid-1980s characterized much of the period under consideration.[26] For an entire postwar generation, gains by the service sector failed to keep pace with industrial losses. Although service sector growth distinguished Philadelphia from the most desperate postwar cities, industrial losses prevented it from experiencing the ongoing prosperity of a city like Toronto.[27] If the 1980s offered some signs that the economic history of the two cities might converge, for decades after World War II this had seemed a remote possibility.

Structures of governance also distinguished between the two urban areas. Early in the postwar era, the government of Ontario set Toronto apart from future rustbelt centers such as Philadelphia. In 1953, when the province created the Metro Toronto government, a federation of Toronto and twelve surrounding municipalities, it linked the fate of the City of Toronto and its surrounding region. In the decades that followed, the provincial government reshuffled the responsibilities of the Metro government—in relation to both its constituent municipalities and the province itself—but regional governance was firmly entrenched. From the mid-1950s onward, Metro took responsibility for key functions including municipal borrowing, public transportation, and regional planning. Over time, it also gained increasing influence in other areas, most notably education, law enforcement, and business licensing.[28] With borrowing backed by the wealth of the entire metropolitan area and the administration of key activities centralized, municipal services were standardized and improved. Perhaps most important—and in contrast to Philadelphia—regional organization enabled a symbiotic relationship between the city core and the suburban periphery. In 1977, John P. Robarts, entrusted to evaluate Metro's record and to make recommendations for its future, explicitly contrasted the Toronto experience to urban development south of the border: "Unlike many large American cities, rapid growth in the suburbs has not brought about a deterioration of the central core." Robarts credited Metro's success to cooperative regional planning and robust public transportation:

> Development policies during the past quarter-century have favoured the concentration of office development in the downtown area and Metro's investment in transit and roads serving the core has supported them. During the same period many downtown industries have relocated in outlying areas on less expensive land, making the central city more desirable as a residential area today than it was some years ago.[29]

Regional government in Toronto was not without its controversies or conflicts. Robarts emphasized the need for political reform to better represent the concerns and interests of Torontonians. However, he identified a crucial difference between the two neighborhoods examined in this study. In Toronto, schools and infrastructure in Little Italy benefited from suburban growth. No similar structure drew suburban wealth back into South Philadelphia.[30]

Partly as a result of its governing structures, Metro became a cohesive urban area. Arteriole streets—maintained by the metropolitan government—with vibrant commercial strips passed seamlessly among Metro's municipalities. The Toronto Transit Commission, also a metropolitan responsibility, carried more riders than any other urban transit system in North America, with the exception of New York.[31] The activities of urbanites also flowed across municipal lines. Residents of the metropolitan area moved widely about the region in their economic and social activities.[32] As the Bureau of Municipal Research put it in 1977:

> *Metro Toronto does function as a single integrated urban area.* In terms of the pattern of travel between home and work, social

> and family networks, ethnic and cultural ties, shopping and entertainment patterns, the use of hospitals, etc., area municipal boundaries have little meaning.[33]

The patterns of social life in Little Italy corroborate this view. Italians moved on a metropolitan scale. The growth and prosperity of the wider region, and its benefits to the urban core, set the stage for Italian ethnic life.

Although the region surrounding the city of Philadelphia was also growing, prosperity on the suburban periphery carried very different implications. The political and fiscal independence of the suburbs meant that economic development on the periphery stripped the city of jobs, population, and tax base. The County of Philadelphia, which might otherwise have provided some regional cohesion, was coextensive with the city. Thus, when wealthier residents and better jobs departed for the suburbs, they left large parts of Philadelphia deprived of resources and services. Although goodly numbers of workers crossed municipal lines in their daily commutes and important entertainment centers drew suburban residents into Philadelphia for recreation, the metropolitan area failed, in crucial ways, to cohere. The most powerful regional dynamic was division. Suburban areas developed shopping, employment, and recreational hubs that increasingly sapped the urban core.[34] These differences in governance brought powerful reverberations in each city, especially given the demographic and economic trends afoot.

The divergent paths of the two urban economies found expression in city space. In the second half of the twentieth century, Philadelphia grew increasingly divided. Like other American cities suffering the brunt of industrial decline, the city divided on racial grounds. Racial segregation resulted from developments common to cities across the United States. Policy decisions at all levels of government; widespread prejudice by employers; the biases of homeowners, real estate agents, and insurers; white flight, and wider economic restructuring all contributed to the racial division of cities in the United States.[35]

Many Philadelphians had a hand in the residential divide. Real estate agents and brokers sought to profitably navigate the city's racial fissure by alternately policing its boundaries and benefiting from its breakdown. Agents from the Commission on Human Relations, created in 1951 to monitor and enforce antidiscrimination laws in housing and employment, found that more than half of the city's brokers and agents either refused African American clients outright, quoted them exorbitant prices in predominantly white areas, or directed them only towards listings in areas already predominantly inhabited by African Americans. At the same time, agents and brokers engaged in "blockbusting," whereby they encouraged panic selling in previously all-white areas once some African Americans had entered. Homeowners in such areas reported receiving hundreds of letters and phone calls warning that their property values stood to plummet and that by remaining in place they risked miscegenation.[36] Although discrimination and blockbusting were in some ways opposite tactics, both encouraged racial division.

Local residents also took up the cause of segregation. African Americans who purchased or rented in predominantly white areas faced violent responses. When Mr. and Mrs. Horshaw moved into the house they had purchased in a largely white section of West Philadelphia in 1960, they were greeted by racist slurs. Over the next several days, stones and bottles filled with ammonia-soaked rags where thrown through their windows, one almost hitting their fifteen-month-old daughter. After several weeks of police protection, the attacks died off and the Horshaws persevered in their new home. When Mr. and Mrs. Figueroa, Puerto Rican immigrants, attempted to move into Northeast Philadelphia, they were met by a crowd throwing rotten eggs. The Figueroas, who intended to rent, elected instead to return to their previous apartment (fig. 2).[37]

Attacks directed against property sent a clear message that opening white areas to African Americans would yield little economic benefit to vendor or purchaser. In the center of Philadelphia, 2608 South Street was twice extensively vandalized prior to African American possession. An African American family purchased it in 1954, but before the new owners could move in local residents ransacked the house, tearing down stair banisters, stripping wall paper, destroying the heating system, and splattering floors with human feces. Four years later, a white owner of the same house rented it to African Americans, whereupon the heating system was again destroyed and racist graffiti scrawled on a wall.[38] These hateful attacks carried personal as well as economic repercussions. Segregationist Philadelphians demonstrated that both individuals and properties would suffer as a result of failures to uphold racial boundaries.

Together, the tactics enforcing residential segregation were highly effective. In 1953, the Commission on Human Relations estimated that 91 percent of the tremendous black population growth was occurring within areas already heavily inhabited by African Americans or on blocks adjacent to them.[39] By 1980, when 40 percent of Philadelphia's population was African American, some 80 percent would have had to relocate to

FIGURE 2 Southwest Philadelphians take to the streets to defend racial boundaries, 1970. The sign reads: "Hey *Whites*! Let's fight for our Rights *Now*." SOURCE: *Philadelphia Evening Bulletin, Photojournalism Collection*, Riots, August 26, 1970. Philadelphia, Temple University Archives, Urban Archives, Philadelphia, PA.

achieve even distribution across the city.[40] Operating throughout Philadelphia, and in relation to such a large portion of the urban population, racial segregation was a reality for all Philadelphians.

Toronto, like Philadelphia, has a history of inequality as well as racial and ethnic tension. From the colonial era forward, black Canadians struggled against pervasive racism in political, legal, and social contexts.[41] Postwar Toronto was no exception. The small black population of the metropolitan area, numbering 0.5 percent of the population in 1971, faced persistent and systematic discrimination in housing, employment, and public spaces.[42] Yet, the small size of the black population meant that prejudices against them played a lesser role in shaping the wider social geography of the city. Meanwhile, larger ethnic and immigrant groups in Toronto clustered into neighborhoods. The most spatially concentrated ethnic group in postwar Toronto was its Jewish population, but Jewish Torontonians suffered little of the economic marginalization experienced by African Americans.[43] Since the 1970s, immigrants to Toronto have come from ever more diverse backgrounds, with Asia, Africa, and the Caribbean contributing significantly to the city's population. While concentrations of new immigrant groups have arisen in the core of the metropolitan area and its suburbs, sociological studies of ethnic enclaves in Toronto have repeatedly rejected parallels between such neighborhoods and African American ghettoization in the United States. Immigrants in Toronto have experienced lower levels of spatial isolation than African Americans in cities such as Philadelphia, and their concentration has been less powerfully interwoven with economic marginalization.[44]

The distribution of poverty in each city at the outset of the 1980s illustrates the contrasting geography of inequality in the two settings. The Canadian census did not designate a poverty line in 1981, but the "incidence of low income" indicates the percentage of households and individuals that spent at least 20 percent more than average on the basic necessities of food, shelter, and clothing.[45] In Philadelphia, the closest comparable statistic is the poverty line, which designates people as poor if they reside in households that earn less than three times the cost of minimal food purchases.[46] While these measures are not perfectly comparable, the geography of economic disadvantage in each city illustrates the distinctive dynamics of inequality in each locale.

At the beginning of the 1980s, both cities housed considerable numbers of economically marginalized people, but the geography of poverty differed. Just over 20 percent of Philadelphians lived in poverty, as compared with approximately 16 percent in metropolitan Toronto.[47] While significant, this difference was less notable than the divergent residential patterns of poor people. By comparison with Toronto, the poor in Philadelphia clustered together, both within census tracts and across larger swaths of the city. In Philadelphia 6 percent of tracts reported poor majorities.[48] More notably, 30 percent of all tracts in Philadelphia were between 25 percent and 50 percent poor. Taken together, the high and moderate poverty areas, areas where at least a quarter of the population lived in poverty, comprised more than a third of the city. Further, the poorest tracts in Philadelphia were grouped together, creating

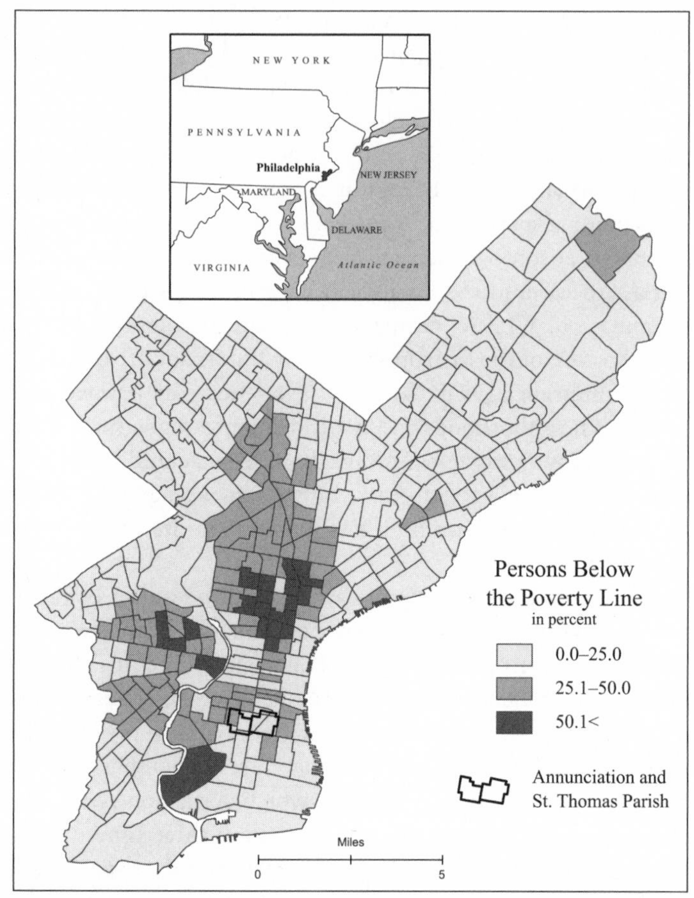

MAP 1 Percent of population below the poverty line by census tract in Philadelphia, 1980. SOURCE: NHGIS, University of Minnesota (1980).

large, continuous areas of poverty. North Philadelphia, the most heavily African American neighborhood in the city, contained the largest of these areas, with six adjacent census tracts reporting poor majorities (map 1). These intensely poor tracts were surrounded by others with moderate levels of poverty; in all, a remarkable 48 adjacent tracts in North Philadelphia reported that more than one quarter of their residents lived without adequate means. West Philadelphia, with a cluster of 20 contiguous census tracts with more than a quarter poor, and South Philadelphia, with a cluster of 12 tracts, also included continuous areas of elevated poverty.

Toronto contained one similar area in 1981, but for the most part the metropolitan landscape was not characterized by large territories of economic disadvantage. The southeast corner of the city center, an area that included the controversial Regent Park project, housed the city's most notable concentration of poor people (map 2).[49] The area contained some of Toronto's worst housing and accommodated some of city's most desperate poor.[50] Outside of this concentration, however, tracts with higher levels of poverty were scattered around the city, surrounded by mixed-income areas. Even Regent Park was bordered on the north by a large wealthy area in which very few residents reported low incomes.

The political, economic, and social history of the two cities had resulted in differing geographies of urban life. The racial division of Philadelphia was also an economic fissure; the postwar era saw African Americans in the city disproportionately concentrated in lower-paying jobs and less-valuable housing.[51] The segregation of African Americans in the city, therefore, also meant the segregation of poverty. Philadelphia was divided into large, continuous territories of economic and racial segregation.[52] Toronto, though never without poverty or inequality, was less distinctly divided. In Toronto, poor families were more likely to live alongside other income groups. No large ethnic group suffered the brunt of focused economic and spatial isolation in the fashion of African Americans in

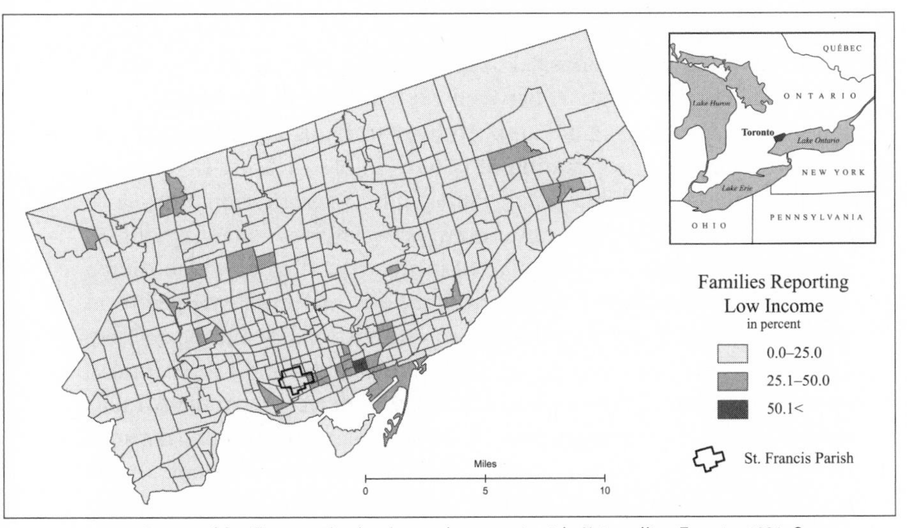

MAP 2 Percent of families reporting low income by census tract in Metropolitan Toronto, 1981. Source: University of Toronto Data Library Service, census tract data (1981). NOTE: This map uses the "economic families" variable (two or more "related individuals" living together—it is the most inclusive Canadian census definition of the family).

Philadelphia. Such differences carried ramifications for the experience of space in each city, differences that were felt in every facet of social life.

The following chapters explore these differences in detail in the lives of Italians, who were among the most visible ethnic groups in both cities. When World War II ended, Italians in Toronto and Philadelphia continued to occupy enclaves that had welcomed immigrants since the nineteenth century. The Italian neighborhood in South Philadelphia stood among the most longstanding in the United States, dating to the mid-1800s.[53] At the center of the original Italian cluster stood St. Mary Magdalen De Pazzi, the first Italian national Roman Catholic church in the country, founded in 1852.[54] In the early years of settlement, Italian groceries and boarding houses accommodated the largely male population of the ethnic enclave and provided an institutional basis for ongoing Italian concentration. Italians in the area knew one another; they moved within common circles. The intimacy of the enclave was evident in the 1856 testimony of Joseph Repelto, who testified in an investigation of alleged voting fraud. As described by Richard Juliani, Repelto knew a good deal about his neighbors, detailing their dates of arrival in the United States, their residences and relocations, and how they earned a living.[55] Tightly clustered, some of them in the neighborhood's first Italian boarding houses, the men were linked by the church and stores they frequented as well as the streets they inhabited. Home, work, local institutions, and common origins interwove to connect Italians living in South Philadelphia.

Indeed, these connections led to the dramatic growth of Italian South Philadelphia in the fifty years between 1870 and 1920. By the later date, almost 64,000 people of Italian birth lived in Philadelphia—making it the second-largest concentration of Italians in North America—and 70 percent of them inhabited South Philadelphia.[56] Institutional development paralleled the demographic boom. Between 1903 and 1917, fifteen new Italian parishes were established in the city, most of which dotted South Philadelphia. In the same period, the city's *padroni* emerged from their small grocery store operations to establish major businesses. Between 1900 and 1929 a single Italian firm brought some 36,000 Italian immigrants to Philadelphia, almost half of whom were from the Abruzzi region.[57] Other immigrants arrived through less formal mechanisms; relatives, friends, and townspeople aided one another in finding a home and work in South Philadelphia. As thousands of Italians poured into the area, rich institutional and social networks emerged.

The two parishes considered here, Annunciation of the Blessed Virgin Mary and St. Thomas Aquinas, stand side by side almost one-third

MAP 3 Annunciation and St. Thomas Parishes, South Philadelphia. SOURCE: ESRI.

of a mile south of the first Italian boarding houses in South Philadelphia (map 3). Both previously dominated by Irish parishioners, the parishes became Italian as the immigrant population swelled beyond its original boundaries. Divided from one another by Broad Street, together they span the fifteen blocks east to west between Sixth and Twenty-first Streets and eight blocks north to south from Federal to McKean Street. By 1940, Annunciation, which became Italian in the first decades of the century, stood at the heart of the enlarged Italian settlement. With Passyunk Avenue, Italian Philadelphia's most important commercial street, passing through the center of the parish, and abundant industrial jobs located just to the east along the Delaware River and north on Washington Avenue, Annunciation developed into a highly concentrated Italian American center.[58] Standing to the west of Broad Street, St. Thomas occupied one of the last areas into which Italian South Philadelphia expanded. As late as the 1920s the church was still predominantly Irish, but by 1940, 96 percent of marriages at the church included at least one spouse with an Italian surname.[59] Unlike Annunciation, St. Thomas stood at the outer boundary of Italian settlement in South Philadelphia. Further from Passyunk and the eastern industrial concentration, St. Thomas was divided at Seventeenth Street between the section heavily occupied by Italians and the section, further west, largely inhabited by non-Catholic African Americans.

Although some Italian immigrants continued to trickle into South

Philadelphia, by the mid-twentieth century sojourners and migrant laborers had been replaced by born-and-bred South Philadelphians. Among those who married at Annunciation and St. Thomas in 1950, the great majority, 87 percent of brides and 73 percent of grooms, had been baptized in one of the churches dotting South Philadelphia. These churches were located close to one another, so by the time of their marriages, most young men and women still lived only a short walk from their places of baptism.[60] In the decades after World War II, the concerns and strategies of Italian South Philadelphians grew out of their longstanding presence in the area. The children and grandchildren of immigrants to the neighborhood, most Italian South Philadelphians had long since ceased to think of themselves as people in motion.[61]

In the postwar decades, however, the stability of South Philadelphia faced novel challenges. Urban decline, racial division, and white depopulation were deeply felt in Annunciation and St. Thomas. Racial tension emerged at familiar flash points. In 1960, African Americans reported virulent resistance to their use of public recreational facilities frequented by white South Philadelphians. Eight years later, longstanding tension surrounding Bok High School—an 88 percent black school on the southern border of Annunciation Parish—irrupted into violent attacks, reprisals, and protests. A public housing project in South Philadelphia, Whitman Park, met with longstanding opposition from Italian South Philadelphians. Racial tension also found expression in chronic skirmishes on city streets as South Philadelphians, most often young men, intermittently attacked one another.[62] As these conflicts simmered, many Italians left the area. The population of Annunciation and eastern St. Thomas fell by more than 50 percent between 1950 and 1990 (fig. 3). Priests in the two parishes reported similar declines in the number of parishioners and schoolchildren under their purview: a combined Catholic population of more than 23,000 in 1950 declined to less than 10,000 by 1990.[63] With wealthier and younger people moving to other parts of the city, the suburbs, and beyond, the parishes were becoming older, poorer, and smaller.

Even as it suffered population loss, South Philadelphia remained a remarkably Italian place. In 1980, when the census first asked Americans about their ethnic "ancestry," more than 10,000 residents of Annunciation Parish reported exclusively Italian origins, and an additional 1,000 residents of the parish listed mixed Italian origins. Together, they represented 75 percent of the inhabitants of the parish. The eastern half of St. Thomas Parish housed almost 3,200 people of exclusive Italian ancestry and some 300 more of mixed Italian ancestry, who

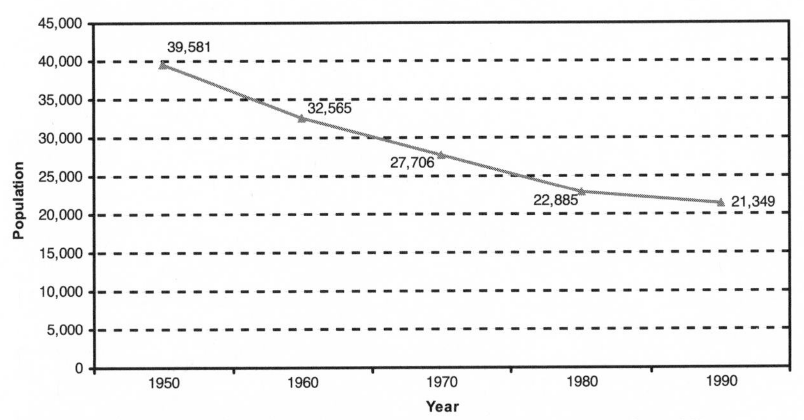

FIGURE 3 Total population of Annunciation Parish and Eastern St. Thomas Parish, Philadelphia, 1950–1990. SOURCES: University of Pennsylvania Library, *Philadelphia Census Tract Data* (1950, 1960) extracted from "The Census Tract Data, 1950: Elizabeth Mullen Bogue File" (National Archives and Records Administration), http://data.library.upenn.edu/phila.html (accessed on March 11, 2005), tracts 1-C, 26-C, and 26-D; NHGIS, University of Minnesota (1970, 1980, 1990), tracts 28-30.

together comprised more than 40 percent of the population. The difference between the core and the periphery of Italian settlement was significant. In Annunciation, Italians comprised a large majority. In St. Thomas they were merely a large minority, only slightly outnumbering the African American population. Nonetheless, both Annunciation and St. Thomas represent an understudied dimension of postwar urban America. Taken together, the two parishes housed almost 15,000 people with Italian origins living in remarkable concentration decades after World War II (fig. 4).[64]

The Italians in Annunciation and St. Thomas Parishes were part of a larger enduring Italian enclave in postwar South Philadelphia. In 1980, over 55,000 Italians in Philadelphia lived as a majority in their own census tracts. Of these, more than 40,000 lived in South Philadelphia, where they comprised more than 80 percent of the population in some census tracts (map 4). The area was home to other Catholic parishes that shared a common history with those at the center of this study. Declining, but still very much Italian, they lingered on.

The postwar history of Annunciation and St. Thomas Parishes promises to reveal much about the larger neighborhood. Annunciation, standing at the center of the wider community, provides an indication of the interior of the neighborhood, the core of the dense social networks in Italian South Philadelphia. St. Thomas highlights another necessary feature of neighborhood: boundaries. The internal division of the parish reflected the wider boundaries of South Philadelphia: the Italians in its

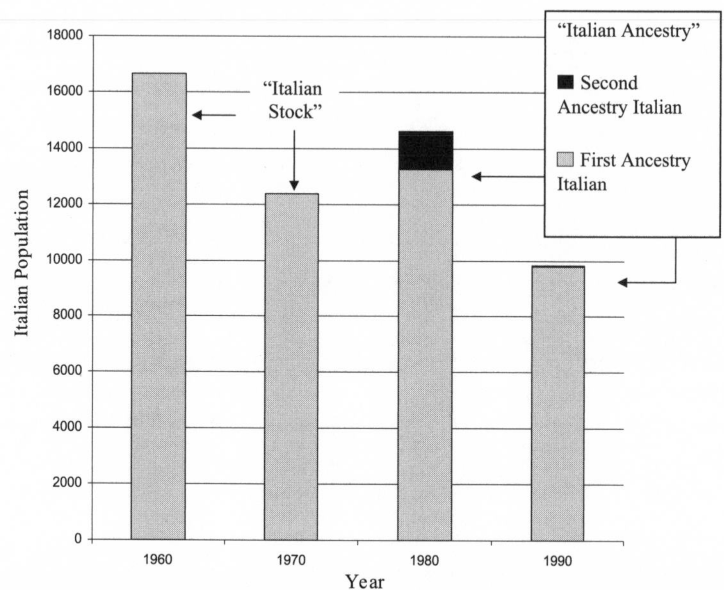

FIGURE 4 Italian Population of Annunciation and Eastern St. Thomas Parish, Philadelphia 1960–1990. NOTE: There is no perfect variable for consideration across all of these census years. In 1960 and 1970, I report "Foreign Stock," a category that includes people of foreign birth as well as people of native birth with foreign-born parents. In the case of mixed foreign parentage, the census provided only the father's place of birth. In the case of mixed native-foreign parentage, the census provided the foreign parentage whatever the sex of that parent. In the later years I use the data for people who reported Italian as their first or second ancestry group. First ancestry group, in light grey, is similar in number—and inclusive of—the people who indicated Italian as their only ancestry group on 1990. SOURCES: University of Pennsylvania Library, *Philadelphia Census Tract Data* (1960) extracted from "The Census Tract Data, 1950: Elizabeth Mullen Bogue File" (National Archives and Records Administration), http://data.library.upenn.edu/phila.html (accessed on March 11, 2005), tracts 1-C, 26-C, and 26-D; NHGIS, University of Minnesota (1970, 1980, 1990), tracts 28-30.

eastern portion occupied the outer edge of Italian concentration, and the African Americans to the west occupied the outskirts of the city's second-largest black neighborhood. The lives of Annunciation and St. Thomas parishioners illuminate the challenges and choices of Italian neighborhoods in the midst of urban crisis.

Italians began arriving to the College Street area in Toronto during the 1890s, when the dilapidated housing available in the "St. John's Ward" no longer sufficed for the incoming Italian population. College Street, to the west of the city core, soon became the city's second Italian residential enclave.[65] As in Philadelphia, local institutions facilitated the beginnings of Italian demographic concentration. Calabrese immigrants, especially from the province of Cosenza, first settled on Mansfield Avenue where one of the area's original Italian residents, Salvatore

Turano, established a grocery store and boarding house. Turano was likely the area's first *padrone*, helping his tenants find work within the city.[66] In 1913 the city's archbishop transferred St. Agnes, at the corner of Dundas and Grace Streets, from Irish to Italian pastoral care, providing institutional basis for an enduring Italian presence in the area.

By the 1920s and 1930s the Italian population of Toronto, though difficult to count due to its transience, had reached 10,000 to 15,000 members, some two-thirds of whom lived in one of three Italian residential enclaves in the city.[67] By the interwar years, the St. Agnes area fostered the kind of intimate neighborly connections that had linked Italian South Philadelphians for decades. A resident of the area in this era later recalled: "In those days people stuck together more. We entertained ourselves; made money . . . we minded our own business in America. We didn't bother with the English."[68] As in Philadelphia, residence, work,

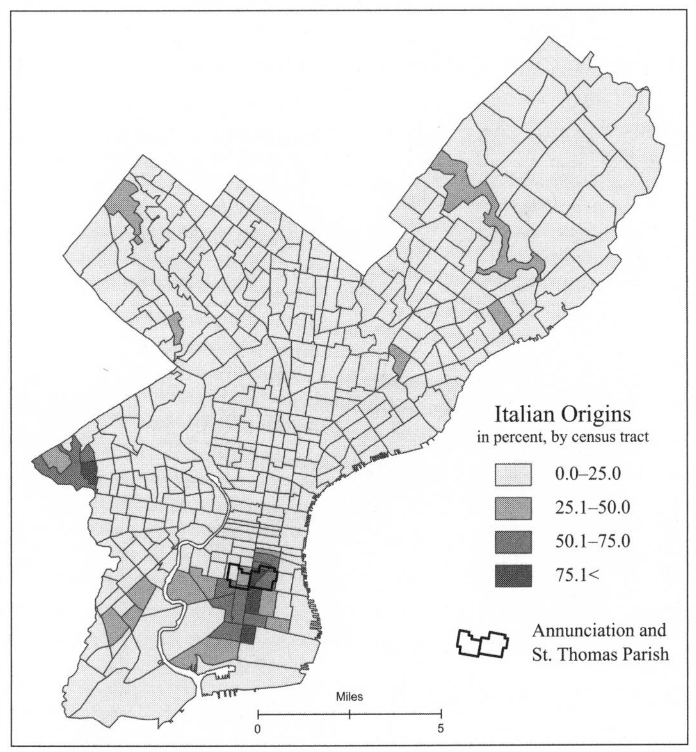

MAP 4 Philadelphians with Italian origins as a percentage of total census tract populations, 1980. SOURCE: NHGIS, University of Minnesota (1980). Calculated using individuals who reported either single or mixed Italian origins.

and leisure combined to associate Italian residents with one another, both in formal institutions, such as the church, and on city streets and backyards. Although Italians never attained the level of local dominance that existed in South Philadelphia during this era, an Italian social world developed.[69] Another resident of the area later described the social activities that connected residents of Little Italy before World War II: "We were playing bocce in the lane there, playing all kinds of games, a lot of picnics we used to have, go to a lot of moonlight dances . . . we used to stick together . . . we respected each other because everybody was poor."[70] A resident of another Italian niche in the city, the declining St. John's Ward, recalled the social connections of Italian Toronto without such wistful nostalgia: "If you married outside the district it was a crime . . . your mother and father had to know the family."[71] Inclusion and exclusion, connection and restriction were the twin dynamics of Italian community—both were rooted in neighborhood.

Unlike South Philadelphia, with more than a dozen Italian parishes, Toronto's Little Italy was always centered in a single parish, first housed in St. Agnes Church at Dundas and Grace, and then a block north in St. Francis Church (map 5). When expropriations and urban renewal displaced the residents of St. John's Ward, St. Agnes and the College Street Little Italy became the oldest surviving Italian parish and residential niche in the city—a bridge between the first and second eras of mass Italian immigration to the city.[72]

MAP 5 St. Agnes/St. Francis Parish, Toronto. SOURCE: University of Toronto Map Library.

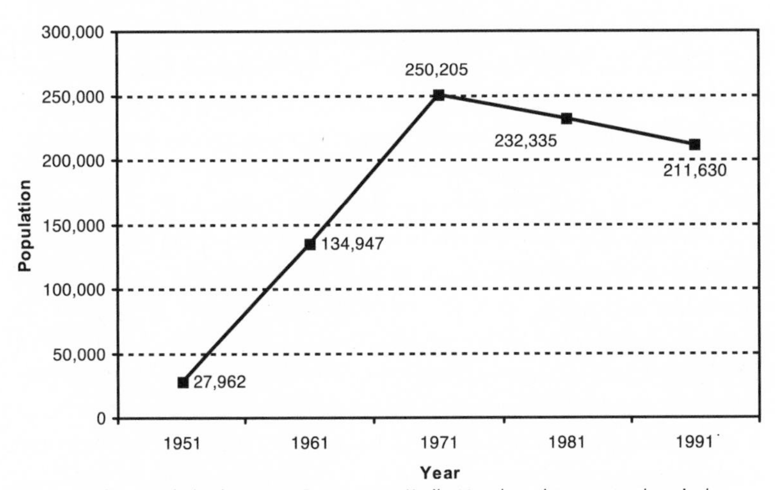

FIGURE 5 Italian population in postwar Toronto. NOTE: Until 1991, these data report only a single ancestry. In 1991, 83 percent of Italian Torontonians reported single Italian origins. SOURCE: University of Toronto Data Library Service (1961–1991).

The Italian population of Toronto was transformed in the postwar era as the preexisting community was numerically overwhelmed by the flood of newcomers. Postwar Italian immigrants shared much with those who had arrived in both Toronto's Little Italy and Italian South Philadelphia in late nineteenth and early twentieth centuries. A significant majority, almost 70 percent, came from southern Italy, where economic underdevelopment—which had prompted mass emigration decades prior—had been exacerbated by the intervening period of immigration restriction in North America, depression, fascism, and war. With the prospects of sustenance from agrarian and artisanal occupations growing increasingly bleak, large numbers of Italians looked abroad for an escape from the specter of poverty. When William Lyon McKenzie King's government removed Italy from the enemy alien list in 1947—as an act of diplomacy and of selective and hesitating efforts to recruit immigrants—Canada emerged as a leading destination for emigrant Italians. In the years that followed, some 20,000 Italians arrived in Canada every year, both to join family members and to serve employment contracts. In 1971, as Italian immigration to Canada subsided, more than 730,000 Canadians reported Italian ethnic origins and more than a third of them lived in metropolitan Toronto (fig. 5).[73]

The timing of Italian immigration, like broader urban development, differentiates the two case studies that follow. By contrast to South Philadelphia, recent arrivals predominate in the story of Toronto's Little

Italy after World War II. Italian-language newspapers, radio, and ultimately television kept immigrants abreast of national and local developments in Italy. Decades after World War II, Italians could still speak their native language, and indeed regional dialects, in institutions and residential enclaves in the city.[74] Although populated by immigrants, the Italian population of Toronto was far from transient. The vast majority of postwar Italian immigrants to Canada, almost 90 percent, chose to remain in their adopted country. For the most part, they had come to Canada in families: men and women came in relatively equal numbers, and adults were often accompanied, or followed, by children.[75] The choices and strategies of Italians in Toronto reflected their enduring presence in the city. Nonetheless, daily life in Toronto provided frequent reminders that many Italians had arrived only recently. The movement of Italians through Toronto, their relationship with urban space, reflected this larger history of motion.

Meanwhile, the presence of immigrants in Toronto—and Canada more broadly—was much discussed by other postwar Canadians. Public debate that often focused on the diversification of Canadian society encouraged Italian ethnics to remain mindful of their origins. Canadian "gatekeepers"—a term that Franca Iacovetta uses for the men and women who "patrolled the nation's entry points" and concerned themselves with the incorporation of immigrants who had already arrived—inserted themselves into ethnic institutions, neighborhoods, and families in hopes of steering immigrants into a narrowly defined Canadian society.[76] Since 1971, when the liberal government under Prime Minister Pierre Elliot Trudeau announced "multiculturalism" as Canada's official stance on diversity, Canadians have debated the impact and significance of a nation founded, at least rhetorically, upon "unity in diversity." However, even those who argue that multiculturalism (in particular as a source of distinction from the United States) has carried only imaginary effects, acknowledge that the policy has placed diverse immigrant origins at the center of Canadian self-conception.[77] For Italian immigrants in Toronto, this was very much the case. The patterns of social life in Toronto's Little Italy reflected the recognition, among Italians, that they were newcomers to the city. Social life in South Philadelphia, by contrast, reflected the rootedness of generations in place.

The new immigrants in Toronto replenished established Italian enclaves—including Little Italy—while simultaneously pouring over old boundaries. In 1961, the presence of new immigrants was already felt in Toronto's Little Italy. At that date, the area housed more than 15,000 people with Italian ancestry, and almost 12,000 of these had themselves

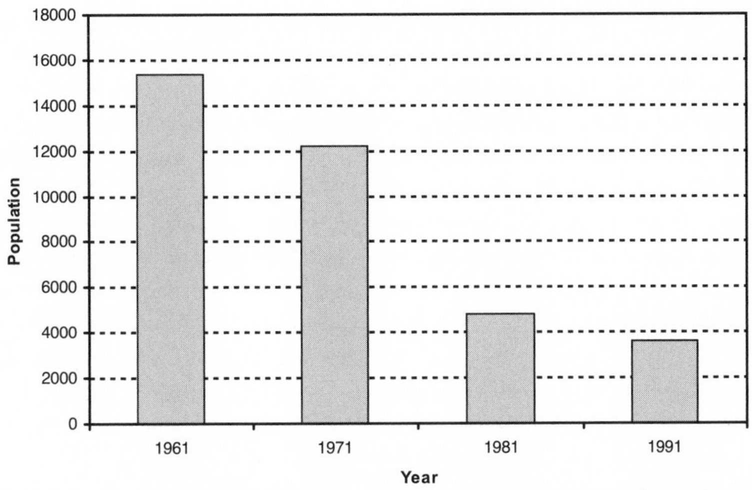

FIGURE 6 Italian origins population of College Street Little Italy, 1961–1991. NOTE: Until 1991, the tract level data report only a single ancestry. In 1991, 90 percent of Italian residents of the parish reported single Italian origins. SOURCE: University of Toronto Data Library Service (1961–1991).

been born in Italy (fig. 6). After a decade of postwar immigration, the area likely housed more Italian immigrants than it ever had in the industrial heyday. In 1971, the area still teemed with immigrants. Of the 12,000 Italians in the College Street area at that date, 8,000 reported Italian birth origins. Yet, a surprising trend was already underway. The Italian population of the area was declining, even as immigrants continued to pour into the city. With the slowing of Italian immigration in the decade that followed, Little Italy became ever less an Italian residential enclave. By 1991 the Italian population of the area had declined to 3,600. The area had never been exclusively Italian—in 1961, 35 percent of area residents reported Italian origins—but by 1991 the falling Italian population amounted to less than 13 percent of the total. A curious transformation had taken place: in the midst of Italian immigration to the city—and immigration to the enclave itself—Little Italy had ceased to be a hub of Italian residential concentration.

The map of Italian Toronto was remade along lines that had been hinted at before World War II. Even before the mass postwar immigration, Italians had begun to settle the corridor to the northwest of Little Italy. In the decades after the war, they poured into these areas. By 1971, the heaviest concentrations of Italians were found in the northwest, along St. Clair and Lawrence Avenues. But even as these other areas supplanted Little Italy, they too became less exclusively Italian. In 1981, some 35,000 Italians lived as a majority in their own census tracts in Toronto, all of them along the northwest corridor. By 1991, the number had dropped sharply to 20,000. The dynamics of Little Italy thus spoke to a wider phenomenon in Toronto.[78] Immigrant influx and concentration was accompanied by simultaneous dispersion.

Shifts within ethnic neighborhoods can only be understood in the wider context of urban change. The period from the 1950s to the 1980s brought contrasting changes to Philadelphia and Toronto. Both cities saw a dispersion of people and economic functions across their wider regions, but these shifts played out very differently in the two contexts. Philadelphia suffered from suburbanization. As people and jobs poured out of the city, they impoverished the neighborhoods that they left behind. Mirroring the divide between the suburbs and the city, Philadelphia itself also splintered along racial and economic lines. Meanwhile, Toronto prospered as a result of suburban growth. Political and economic developments preserved a connection between the city and the wider region, and homes in older city neighborhoods continued to grow more valuable. The social history of immigrant reception—never as violent or contentious as African American settlement in cities such as Philadelphia—prevented the dramatic division of the urban landscape. Wider urban processes reshaped neighborhoods such as Italian South Philadelphia and Toronto's Little Italy.

2 Italian Markets: Real Estate Exchange and Ethnic Community

In its 1965 hearings on racial discrimination in the housing market, the Philadelphia Commission on Human Relations permitted Noel Smorto, a South Philadelphia realtor, to dwell at length on the "unique" situation in the area where he lived and worked:

> There is a certain group of people, particularly the Italian people, who want to live in this area. They take a great deal of pride in their homes, and they are very jealous as to who moves into these areas. They build friendships, relationships. They have their families close by . . . [they] have put all their life's savings, their money, and worked hard to improve a house, and they expect to remain there in their remaining years in comfort and in close proximity to their friends and relatives.[1]

For Italian South Philadelphians, the payoff for a lifetime's investment came not with resale, but rather with the passage of years within rich social networks. A house in the old neighborhood held distinct social value. The "jealousy" of local Italian residents was particularly piqued when this vision was threatened by African American would-be buyers. Faced with a potential African American purchaser, Italian South Philadelphians used their close relationships to exert coercive pressure. Smorto, who did business from a South Philadelphia office at Seventeenth and Passyunk, reported that the menacing side of social networks emerged when he showed a local home to African Americans: "I got calls at home. I was threatened—my neighbors, my home, family. I had people call me, pleaded with me . . . 'Please protect us; keep them out.'"[2]

Had he visited Toronto's Little Italy, Smorto might have noticed that there too the housing market had a surprisingly social character. The neighborhood ties of South Philadelphia represented but one way that ethnicity could shape a housing market. John D., who got his start as a real estate agent in Toronto's Little Italy in the 1970s, later reflected on the Italian approach to home ownership in that area:

> [M]y own clients and myself, we bought originally . . . number one for pride of ownership . . . you feel different if you own something . . . but number two because . . . the property will go up in value . . . we want to establish a pattern . . . where we can use equity and the appreciation to buy a better house or to do something else with the money.

For the residents of Toronto's Little Italy, houses were crucial economic assets. Immigrant families who shared ownership of a home in Little Italy with other recent arrivals could use the accrual of equity to purchase their own single-family homes. Others used their growing economic resources to buy larger houses elsewhere in the city. John also recalled home owners who remained in place but withdrew equity from their homes to "take a trip around the world." While ownership brought a sense of satisfaction, in Toronto's Little Italy it was also tied to economic strategy. In John's view, "That's why it's very important to own a piece of real estate."[3] Italians in Toronto took advantage of rising property values with help from one another. The postwar housing market in Toronto was characterized by high prices and profitable exchanges, but until the late 1960s Canadian mortgage regulations discouraged bank loans to purchasers of older inner-city housing. Even as regulations relaxed, credit remained scarce in places like Little Italy. For their part, immigrants were sometimes reluctant to use banks, with their restrictive terms and demanding application processes.[4] As a result of the scarcity of credit and their own preferences, Italian city residents generated their own credit. Vendor "take-back" mortgages, wherein the buyer obtained a mortgage from the seller, assumed dominance in Little Italy. In this context, Italians disproportionately sold to other Italians; their personal loans were arranged within ethnic social networks. The spatial character of such bonds differed from those in South Philadelphia. Rather than linking Italian ethnics in local spatial control, the exchange of properties in Little Italy linked people spread widely across the city. Sellers, lenders, and agents active in the housing market of Little Italy were often geographic outsiders but ethnic insiders, reflecting the elasticity of Italian social networks in Toronto.

Housing markets play an integral role in defining the relationship between ethnicity and urban space. Often as a result of market dynamics outside of their control, residents of Italian South Philadelphia and Toronto's Little Italy had different experiences of housing. Neighborhoods like the one described by Smorto—where locality and social connection overlap—are facilitated by particular market conditions. In the stagnant market of Italian South Philadelphia, residents valued their houses primarily for their use. In Toronto's Little Italy, rapidly rising housing prices encouraged alternative forms of social connection. Properties in the booming market of Toronto's Little Italy changed hands frequently, and residents found social connection outside of their immediate vicinity. Focusing on Catholic parishes at the core of each Italian neighborhood, this chapter uses deeds of sale to detail the constraints and opportunities available in each housing market.[5] The very different stages set by the two markets encouraged Italians in each place to make distinctive use of ethnic bonds.

Even as real estate markets shaped ethnicity, housing exchanges were the product of ethnic relations. Ethnic social ties permeated both housing markets at many levels. By linking sellers to buyers, lenders to borrowers, and agents and lawyers to clients, Italian social bonds facilitated real estate transactions. In Italian South Philadelphia, the neighborhood basis of social life and concerns about racial turnover steered housing transactions. In Toronto's Little Italy, the efforts of immigrants to achieve and preserve individual economic stability shaped market choices. The second part of this chapter uses deeds, mortgages, and housing advertisements to explore the spatial contours of ethnicity within each real estate market. Ethnicity and urban economies interwove in the creation of two very different forms of city life.

On Saturday, May 7, 1960, the *Toronto Star* ran an advertisement for a house in Toronto's Little Italy: "$1,000 down Bathurst College . . . house solid brick new furnace hardwood floors throughout move right in."[6] The description might have suited one of the larger dwellings in either Italian residential niche. Viewed from the perspective of bricks and mortar, flooring and furnaces, the houses in Little Italy and Italian South

Philadelphia greatly resembled one another. Although some houses in each Italian area included storefronts, garages, and outbuildings while others did not, they were for the most part modest homes of working urban dwellers: compact, simple, and homogeneous (figs. 7 and 8).

Both South Philadelphia and Toronto's Little Italy had been built well before World War II, and little new housing construction occurred in either area in the ensuing decades. The two- and three-story row houses in Annunciation and St. Thomas Parishes in Philadelphia date from the late nineteenth century, and in the years that followed most underwent only modest alteration. Although residents subdivided rooms, built additions, and resurfaced their houses in brick, permastone, stucco, vinyl, and aluminum to accommodate their changing needs and tastes, the houses remained substantially unchanged.[7]

The houses in Toronto's Little Italy differed only slightly from those of South Philadelphia. With few exceptions, the two- and three-story buildings in the area—single-family homes as well as small multiunit structures—had lined the streets of Little Italy since the late nineteenth and early twentieth centuries. As in Philadelphia, many houses under-

FIGURE 7 Row homes at Thirteenth and McKean in Annunciation Parish, South Philadelphia. SOURCE: *Philadelphia Evening Bulletin, Photojournalism Collection,* Streets-McKean, June 14, 1975. Philadelphia, Temple University Archives, Urban Archives, Philadelphia, PA.

FIGURE 8 Houses on Manning Street in Toronto's Little Italy. SOURCE: 250 Manning (houses opposite school site), City of Toronto Archives, series 840, file 348, folio 17 (1959).

went minor alteration—residents made changes and repairs, often on the inside rather than the exterior of their houses.[8] Perhaps the most significant physical difference between the houses of Italian South Philadelphia and Toronto's Little Italy was the size of the lots on which they were situated. Lots in both areas were relatively narrow—ranging between 10 and 20 feet of street front—but in Toronto they stretched much further back from the street, often providing residents with both back and front yards. In Italian South Philadelphia, lots averaged half the length of their Toronto counterparts; the open space behind houses was small, and front yards were rare.[9] Residents of Little Italy thus benefited from the sunlight, ventilation, and open space missing in South Philadelphia.

The advertisement for the Bathurst Street home did include further details that distinguished it from properties in South Philadelphia, but they did not touch upon acreage or sunlight. Noting that the house included two kitchens, the ad suggested that it was a "terrific income house." The house was deemed a potential earner because of its location in Little Italy. Its two kitchens accommodated the settlement strategies of immigrant Italians pouring into the area.

On sampled streets at the core of Toronto's Little Italy, the great majority of homes were occupied by their owners, but many owners rented either rooms or apartments to boarders and tenants. The number of houses with renters increased steadily in the postwar era from almost 40 percent in 1950 to a high of 56 percent in 1970.[10] Italian immigrant recollections of this period frequently include time spent renting quarters in the home of another family.[11] Tenants and owners living in close proximity sometimes found the arrangement invasive. One Italian immigrant to Toronto in 1966 recalled that renting a room in her landlord's house involved constant infringement of her privacy and autonomy; her landlords imposed "so strict rules that you couldn't even move in that house . . . I had to watch when I could open the water, I couldn't take a bath every day." Nonetheless, she, like many other immigrants, found herself with few alternatives in the period immediately after arriving in the city: "That was the situation not just for me but all the people who was rented. There was no apartments at that time."[12] Also during the mid-1960s, Vince P.'s family found space in a three-story house on Grace Street alongside four other families. There too, the bathroom epitomized the crowded conditions: "You can well imagine the morning . . . the line up for the bathroom. These kids, right, everybody's going to school . . . it was kind of pandemonium."[13] For some, cramped housing in Little Italy resembled premigration life. In the 1960s, Paul N.'s family shared a three-bedroom house on Richmond Street with the families of his two maternal aunts. He reflects: "It didn't make a difference to me . . . I mean in Italy there wasn't three or four bedrooms . . . a little corner, that's where your bed was . . . I didn't even think about it."[14]

Shared space allowed new immigrants to limit their expenses while getting settled in Canada, but the rental income was a particular benefit to homeowners. Many Italian immigrants purchased homes financed, in part, by rental income. As late as 1990 almost half of the homeowners on sampled streets in the College Street area rented rooms in their residences. Such rentals were rare in Italian South Philadelphia, where a combination of low immigration and cheaper housing allowed most homeowners in Annunciation and St. Thomas to occupy their houses without renters or boarders.

Little Italy's houses could also be considered good investments because property values in the area were rapidly rising. Prices in Italian South Philadelphia and Toronto's Little Italy may have differed somewhat in the 1940s before systematic data became available for the former. However, a dramatic divergence in prices emerged between 1950 and 1990 as the real value of houses rose dramatically in Toronto's Little Italy but languished in Italian South Philadelphia. This growing difference encouraged increasingly divergent understandings of the role and use of housing.

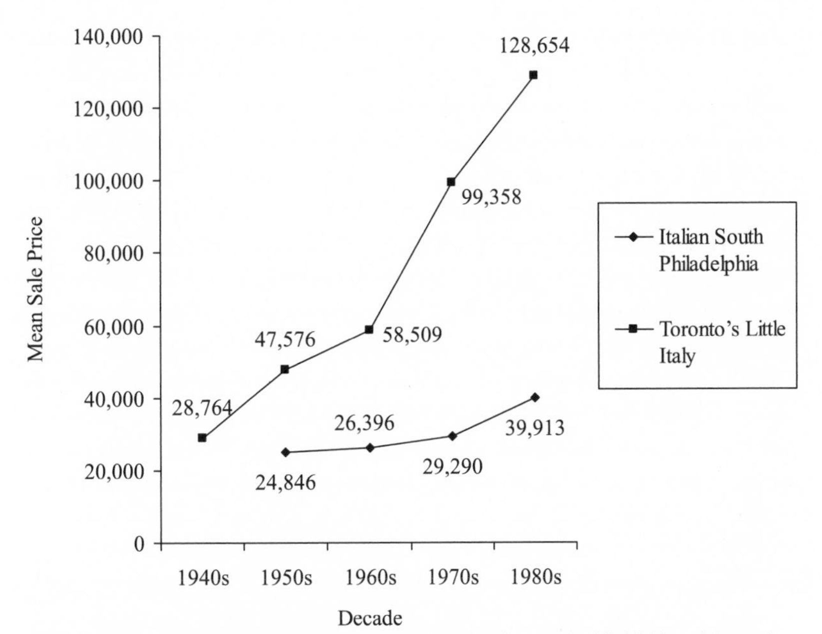

FIGURE 9 Mean price of properties sold for more than $10 in Italian South Philadelphia and Toronto's Little Italy, 1940–1990 (USD 1989). SOURCE: Philadelphia Deeds of Sale, Summary Sheets, 1950–1990, Office of Records, City Hall, Philadelphia; Toronto Deeds of Sale, 1940–1990, Office of Land Title/Land Registry, Ontario Ministry of Consumer Affairs, Toronto. Deeds of sale for 300 randomly selected properties in Philadelphia and for 150 properties in Toronto. *N* = 763 (Philadelphia 275, Toronto 488).

In Toronto's Little Italy, homeowners saw a remarkable quadrupling of values between the 1940s and the 1980s (fig. 9). Houses that sold for less than $30,000 in the 1940s fetched upward of $128,000 by the 1980s (all prices and earnings are reported in 1989 USD). The rapid rise in prices offered tremendous profit to shrewd investors. Sam and Josie Demacio, for example, earned more than $28,000 by purchasing a house at 313 Manning in 1951 and selling just a year later. This trend continued in the decades that followed. Leonardo and Vincenza De Felice purchased 318 Manning for $56,000 in 1972, a year that would see some of the largest increases in real estate values in Toronto history, and sold it the next year for $112,800. Perhaps most remarkable were the properties with few sales during these years. The Stefano family purchased 228 Claremont in 1948 for a price of $13,822; when they sold the house to Shou and Liang Yu in 1989, it fetched $180,600. Although the booming Toronto market did not prove lucky for all investors, most housing purchases, whether for rapid resale or for long-term use, rewarded investment. Italian immigrants arrived in Little Italy while property values were climbing and they profited from the rise.

In Italian South Philadelphia, meanwhile, most homeowners did not experience a similar boom. Between the 1950s and the 1970s prices remained low as increases from one decade to the next stood at less than one-fifth of comparable figures in Toronto (fig. 9 above). The stagnation of South Philadelphia's housing market was especially acute in St. Thomas Parish, on the outskirts of the Italian residential area. St. Thomas actually experienced an absolute decrease in prices from the 1950s to the 1970s. In the 1980s, as the economy of Philadelphia revived, the real estate market in Italian South Philadelphia improved, with the prices increasing more than 30 percent. Lying at the center of Italian South Philadelphia, Annunciation saw particular price growth (fig. 10). After its significant gains of the 1980s, the market in Annunciation came to resemble that of Toronto's Little Italy, with some owners taking advantage of rapid resale. For example, Joseph and Margaret

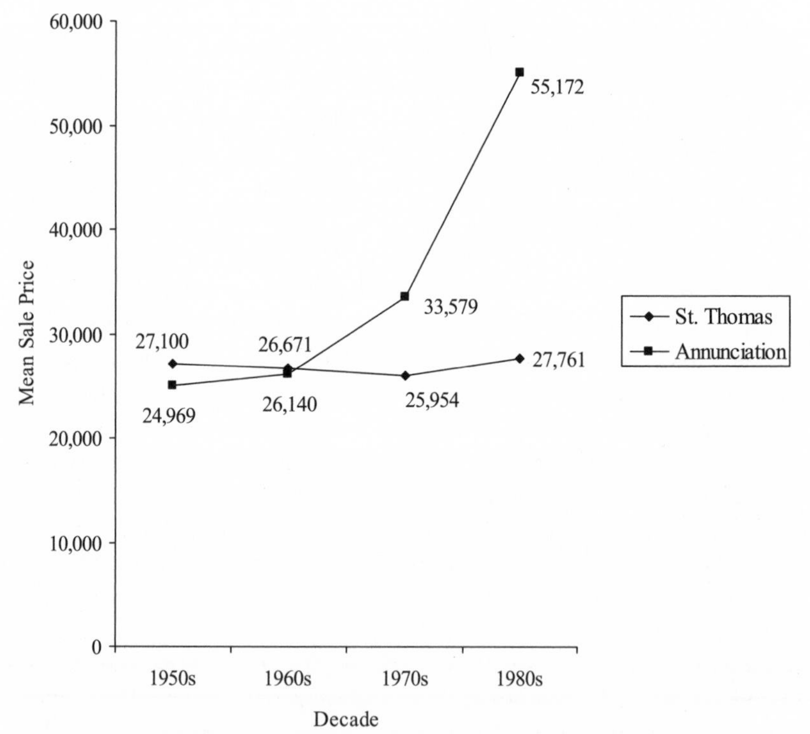

FIGURE 10 Mean price of properties sold for more than $10 in Italian South Philadelphia, by parish, 1950–1990 (USD 1989). SOURCE: Philadelphia Deeds of Sale, Summary Sheets, 1950–1990, Office of Records, City Hall, Philadelphia; Toronto Deeds of Sale, 1940–1990, Office of Land Title/Land Registry, Ontario Ministry of Consumer Affairs, Toronto. Deeds of sale for 300 randomly selected properties in Philadelphia and for 150 properties in Toronto. *N* = 763 (Philadelphia 275, Toronto 488).

Tarragrossa netted $15,000 by purchasing 1024 Tasker Street in 1982 and reselling four years later. However, until the 1980s Annunciation bore greater similarity to St. Thomas, where homeowners understood that the value of their houses increased slowly, if at all.

Divergent experiences of housing in Italian South Philadelphia and Toronto's Little Italy reflected the wider real estate markets in the two cities. As noted in chapter 1, housing values in Philadelphia stagnated in the postwar period before beginning to rise in the 1970s and 1980s.[15] Price increases may have come somewhat later in South Philadelphia than in the city as a whole, but broadly speaking, local real estate followed citywide trends.[16] Until the 1980s, Italian South Philadelphians could expect little, if any, financial return on their investments in housing.

In Toronto, citywide real estate values increased steadily throughout the postwar period. Like other Torontonians, residents of Toronto's Little Italy could view their homes as sound investments, likely to appreciate in value over time. This appreciation did not mean that Little Italy's homeowners had grown wealthy in comparison with other Torontonians. Despite the appreciation of property values, houses in Little Italy sold for half as much as those in the metropolitan area overall. Nonetheless, most Torontonians in any postwar decade could view homeownership as a potential vehicle of economic prosperity. By selling a Little Italy home to purchase another, often larger, house elsewhere in the city, Italian Torontonians could combine the economic fruits of property investment with their other resources as they built their personal prosperity.[17]

Although Italian South Philadelphians could expect little economic profit from housing, their demographic dominance within the parishes meant they could realistically hope to retain an ethnically homogeneous neighborhood. By contrast, despite ongoing immigration, Italians in postwar Little Italy comprised only the largest minority of local residents; they never commanded the large majorities achieved in Italian South Philadelphia.[18] Correspondingly, people of Italian extraction controlled most local residential real estate in Italian South Philadelphia. In Annunciation and the eastern section of St. Thomas Parish, a large majority, never less than two-thirds, of sampled sellers bore Italian surnames.[19] In Toronto's Little Italy, Italian ethnics dominated far less, never reaching two-thirds of sampled grantors. The presence of Italians among sellers of real estate rose steadily during the postwar Italian immigration to Toronto, peaked in the 1970s, and declined in the fol-

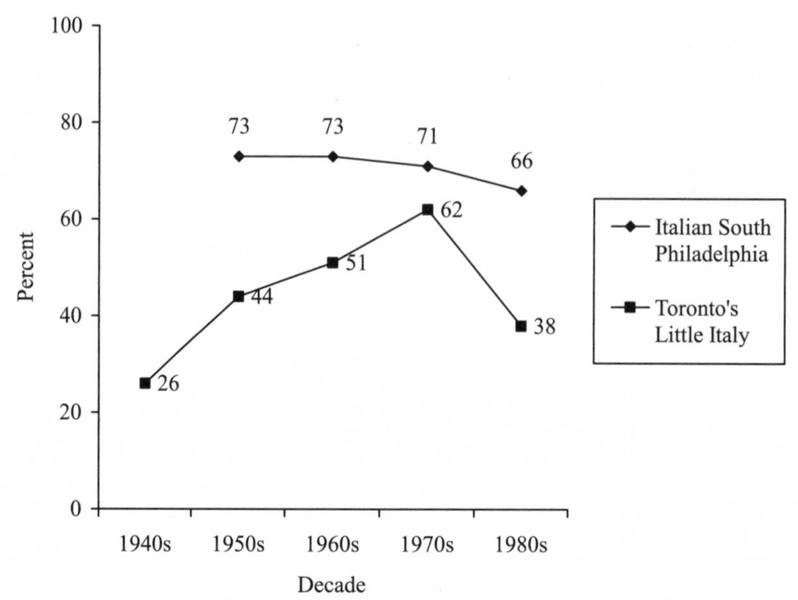

FIGURE 11 Percentage of sellers with Italian surnames, Italian South Philadelphia and Toronto's Little Italy, 1940–1990. SOURCE: Philadelphia Deeds of Sale, Summary Sheets, 1950–1990, Office of Records, City Hall, Philadelphia; Toronto Deeds of Sale, 1940–1990, Office of Land Title/Land Registry, Ontario Ministry of Consumer Affairs, Toronto. *N* = 889 (Philadelphia 358, Toronto 531).

lowing decade (fig. 11). Even at their apex, Italian Torontonians never controlled the real estate of Toronto's Little Italy to the extent of their counterparts in South Philadelphia.

Distinct demographic and economic contexts fostered different expectations of housing. In Italian South Philadelphia, residents might well expect to determine who entered their neighborhood because pressure exerted within Italian social networks could control the residential population. Italians in Toronto's Little Italy could never hope for similar control of their neighborhood. Ongoing immigration, residential flux, and ethnic diversity conspired against territorial dominance.

Italian homeowners were constrained by their economic and demographic circumstances, but they fashioned their own local responses. In Italian South Philadelphia, residents responded by selling property relatively rarely. Among the properties that sold at least once between 1950 and 1990 in Italian South Philadelphia, almost 80 percent sold on only that one occasion or just once again at some later date.[20] Thus in a forty-year period, the great majority of houses in Italian South Philadelphia were inhabited by only two or three different owners (fig. 12). This strategy meant that the vision conjured by Noel Smorto could be realized, at least in part. Life on South Philadelphia streets was characterized by the long-term persistence of neighbors. On Bancroft Street in St. Thomas Parish, four Italian families lived side by side for the entire period from 1939 to 1960.[21] Overall, 50 percent of residents on sampled streets in South Philadelphia in 1980 had lived in the same house for ten years or more, 20 percent for at least thirty years, and 10 percent had remained in place for the entire period from 1950 to 1990.[22]

Some South Philadelphia families and individuals remained in place even longer than these figures suggest. Cathy S. was born in 1944 in a bedroom on the third floor of her family's home in Annunciation Parish. Her father and his parents had resided in the house since the early 1920s, joined in 1943 by Cathy's mother after a wedding at Annunciation. Thirty years later, Cathy and her newlywed husband moved

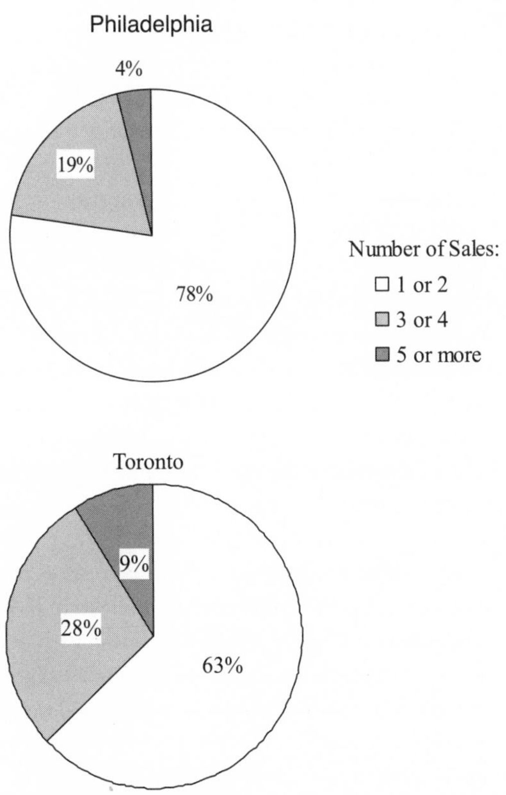

FIGURE 12 Percent of properties with selected number of exchanges, Italian South Philadelphia and Toronto's Little Italy, 1950–1990. SOURCE: Philadelphia Deeds of Sale, Summary Sheets, 1950–1990, Office of Records, City Hall, Philadelphia; Toronto Deeds of Sale, 1940–1990, Office of Land Title/Land Registry, Ontario Ministry of Consumer Affairs, Toronto. *N* = 889 (Philadelphia 358, Toronto 531).

into her grandparents' former bedroom on the second floor, where the couple resided until his death in the late 1990s. In 2007, still sleeping one floor beneath her birthplace, Cathy attempted to describe the experience of remaining in one place for a lifetime: "It's home. It's the only home you've ever known. I could walk around in my house with all the lights out and not bump into anything."[23] Ida G. has resided in her house in St. Thomas Parish for her entire eighty-seven years. She cannot recall ever considering relocation: "Where were you going to go? . . . I had some friends . . . I had nice neighbors . . . where would I go?" After staying in place for so long, she reflected, "you don't realize what's on the outside."[24]

In the heated real estate market of Toronto's Little Italy, local persistence was more unusual. Only 63 percent of homes changed hands twice or less between 1950 and 1990, and almost 10 percent changed hands at least five times (fig. 12 above). These patterns of sale carried ramifications on city streets. In 1980, 38 percent of residents had lived in the same house for at least ten years, 7 percent for at least thirty years, and only 5 percent lasted the entire period from 1950 to 1990.[25] The real estate practices of Toronto's Little Italy encouraged greater residential flux—few people grew old in the company of longtime neighbors.

Italian Torontonians, many of them immigrants or the children of immigrants, understood housing exchanges and residential relocations within larger narratives of international migration. Considered in light of migration from Italy, a move across the metropolitan area seemed a comparatively small dislocation. Paul N., who owned his first home on Dundas Street in Little Italy for only a few years in the late 1960s before selling and moving to the suburbs, saw this decision in the context of his mother's choice to leave her home in Puglia. In his view, a multigenerational struggle for economic security took precedence over attachment to particular places:

> When you're young you have to think about making a life for yourself . . . if it means moving from Italy to Canada, which my mother did, did she have a lack of loyalty for Italy? I don't know, I don't think so . . . We left Dundas to go to Port Credit, was there a lack of loyalty to community? I don't know . . . my loyalty was to my family.[26]

In this case, the resale of Paul's property in Little Italy netted some $13,000, profit that he used to purchase a larger home for his growing family. Vince P., whose family immigrated from Calabria in the early

1960s, purchased his first home in 1974, after his marriage. He too recalls the economic benefits of purchasing housing outside of Little Italy: "What determined that [relocation] was economics, the further north you went the more affordable houses were . . . so we settled along Eglinton . . . Eglinton and Dufferin."[27] Residents of Toronto's Little Italy thought of housing in financial terms. While they surely took pride in their homes and valued their neighbors (like their peers in South Philadelphia), immigrant families approached property exchange as part of a larger economic strategy. Residential persistence within a single neighborhood, "loyalty" to Little Italy, had little place within this outlook.

Not all residents of the area chose to depart for economic gain. For some, the advantages of residing in Toronto's Little Italy outweighed the temptation of larger houses in newer neighborhoods. John D. found that some clients resisted the temptation to sell their homes despite decades of encouragement. "Next time, next time," they would suggest. Of course, with property values rising in Little Italy, even those who remained in place could view their decision as economically prudent. Asked if he regrets having left the area in the 1970s, Vince P. reflects, "Not really, although I know I can't afford to come back now. If you had real estate here now it's worth a lot."[28]

The dynamics of each local real estate market encouraged different kinds of social bonds. Ethnic homogeneity and persistence in Italian South Philadelphia were facilitated by the depressed conditions of the real estate market. Homeowners had relatively little incentive to sell, especially given their refusal of African American buyers. Further, although many young people departed South Philadelphia for the suburbs and other parts of the city, others could afford to purchase homes close to their parents. Their presence in South Philadelphia cemented local social solidarity. Hence, the expectation of Italian South Philadelphians to live surrounded by friends and family was encouraged by the local housing market; economic conditions fostered stable, local social ties. In Toronto, the real estate market encouraged more frequent sales and greater geographic fluidity. Ethnic life was shaped by the expectation that properties purchased would be good earners. Italian residents were less able and less inclined to establish stable local ties, and Italian ethnicity developed without the same degree of local cohesion.[29]

Just as divergent housing markets set the stage for different social experiences, social life impinged upon the market. Urban decline and prosperity, and the housing price trends that they entail, are not merely variables in urban experience—present or absent in respective contexts—but historical processes that develop over time in the daily activities of urban residents. If prices in Italian South Philadelphia fell because of events in the rest of the city, the responses of Italian residents, and their use of Italian ethnicity to navigate the declining market, were entirely their own. Similarly, although prices rose in Toronto's Little Italy because of citywide prosperity, Italian Torontonians crafted their own real estate networks.

Viewed as social documents, real estate records divulge a great deal about the operation of ethnicity within housing markets. The names and addresses recorded in deeds of sale, mortgages, and real estate advertisements allow analysis of much more than price trends. The exchange of a house requires an encounter of buyer and seller; people come together to transfer properties. The documentary traces of such encounters go a long way in detailing a social history of each housing market. Real estate records illuminate the social and spatial experiences of stalled and prosperous real estate markets and offer insight into how concerns about race, space, and economy shaped postwar Italian ethnicity.

Despite the differences between the two housing markets, Italian sellers in both areas showed strong preference for selling their homes to other people of Italian extraction. The real estate market in Toronto's Little Italy already exhibited substantial ethnic segmentation in the 1940s. In that decade, most sellers were non-Italian, and they sold to an Italian only one in four times, whereas sellers with Italian surnames sold to other Italians in three out of four cases (fig. 13). Perhaps the tension arising from Italy's position in World War II and the internment of a number of Italian Torontonians—including the priest at the church in Little Italy—caused a particularly deep divide between Italians and non-Italians during this decade.[30]

Despite the particularities of wartime Toronto, the ethnic division of the real estate market in the 1940s was part of a larger pattern. Italians in both areas showed a strong tendency to sell to other Italians throughout the postwar era. Between 1940 and 1990, more than 60 percent of Italians sold to other Italians in both areas, and the figure in some decades rose above 80 percent. In contrast, non-Italians sold to Italians in only 33 percent of cases.

The overrepresentation of Italian buyers when the sellers were also of Italian origins points to the powerful role of ethnic networks in the real estate markets of each city. Italian vendors were especially likely to find Italian buyers because of some further social connections. In Italian South Philadelphia, transactions were often an outgrowth of neighborhood social ties and often did not involve a realtor. Anthony S. purchased his home in Annunciation Parish from a customer to whom

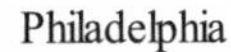

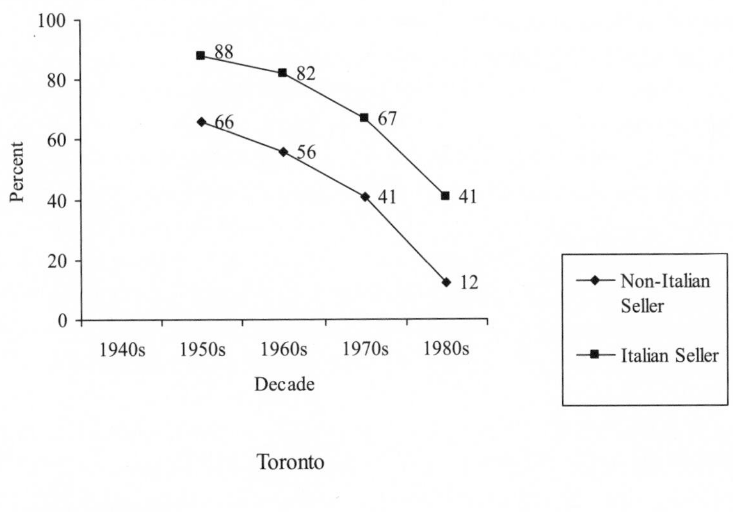

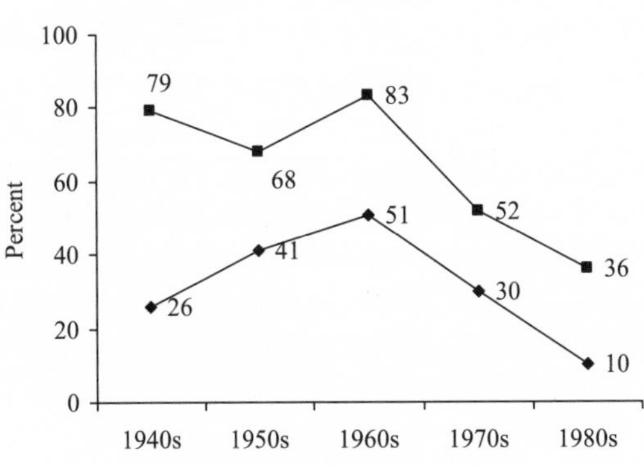

FIGURE 13 Sales to Italian buyers as a percentage of total exchanges, by seller's ethnicity, Italian South Philadelphia and Toronto's Little Italy, 1940–1990. SOURCE: Philadelphia Deeds of Sale, Summary Sheets, 1950–1990, Office of Records, City Hall, Philadelphia; Toronto Deeds of Sale, 1940–1990, Office of Land Title/Land Registry, Ontario Ministry of Consumer Affairs, Toronto. $N = 889$ (Philadelphia 358, Toronto 531).

he delivered alcohol every Saturday as part of his neighborhood "beer route." As a personal favor to this particular customer—who "was old and couldn't see too good"—Anthony would carry the delivery downstairs to the cellar refrigerator and leave with the previous week's empty bottles in tow. "I did this for a year," Anthony recalls, "and I kept telling him . . . 'You know this house is a nice house, got a big yard . . . when you get rid of this house you gotta tell me.'" When the customer eventually decided to sell, he called Anthony and his wife, announcing:

"I tell you one price, no up no down. You like, you buy, you no like, goodbye."[31] The couple accepted the price and soon began moving in. Cynthia C.'s family, also in Annunciation Parish, moved across their street in 1967, purchasing a new home from "good friends." In Cynthia's case, "we knew that they were retiring and moving way up in Bucks County. And there was not even a sale sign. We heard it and my mom said, 'Let's talk.' And agreed on a price and that was that."[32]

Cynthia, who also observed real estate markets as a mortgage banker for several decades, describes sales between acquaintances as commonplace in South Philadelphia until the rise in prices in the 1980s. In the 1960s and 1970s, "people were happy to sell the house, and *happy* to sell the house to somebody they knew. Especially if they loved the neighborhood and they loved the house . . . it was almost like following through with family." With the rise in prices, priorities changed: "Today with the values the way they increased . . . people are looking for the top dollar."[33] As South Philadelphians looked outside their circles of friends and acquaintances for purchasers, they grew more likely to sell to non-Italians. In the 1980s, Italian sellers grew dramatically less likely to sell to Italian buyers, with the figure plummeting from almost 70 percent to 40 percent (fig. 13 above).

Sometimes, sales between coethnics quite literally "followed through with family"—the housing market in Italian South Philadelphia included a remarkable number of exchanges between people with the same surnames. In Annunciation and St. Thomas, 17 percent of all sampled sales transferred property between nonspouses with the same last name.[34] In the great majority of these cases, 80 percent, the property exchanged for less than $10. The properties were gifts, most often to the grantor's children. In addition, a large number of properties were exchanged for $10 or less by people who did not share names. Most likely, many of these transactions still involved parents and their children, especially married daughters receiving property from their parents. Italian owners were especially likely to give their properties as gifts, doing so 26 percent of the time, twice as often as their non-Italian counterparts (fig. 14). Gift exchanges represented a major segment of the market in every decade, reaching, at their height in the 1970s, more than a third of sales by Italians. Most likely, many parents expected some form of repayment for a gift house, but such exchanges nonetheless kept the houses off the market and within the family. Added to those that were inherited, gift exchanges in Italian South Philadelphia comprised a tremendous proportion of the total market.

The frequency of gift exchanges in South Philadelphia simultane-

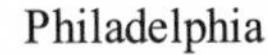

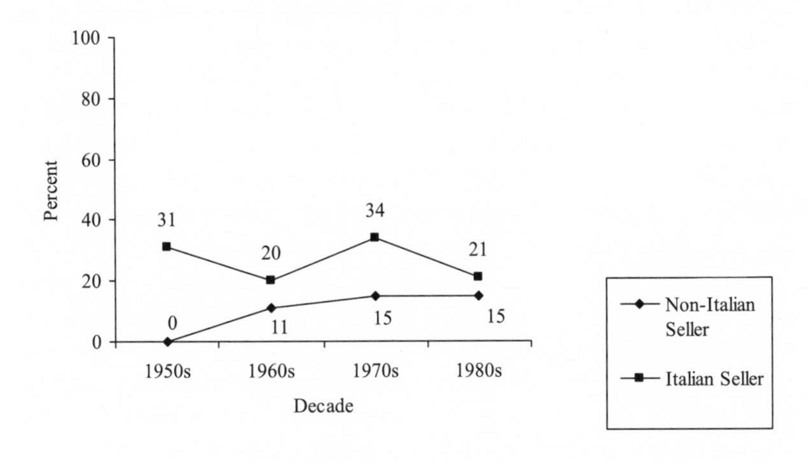

Toronto

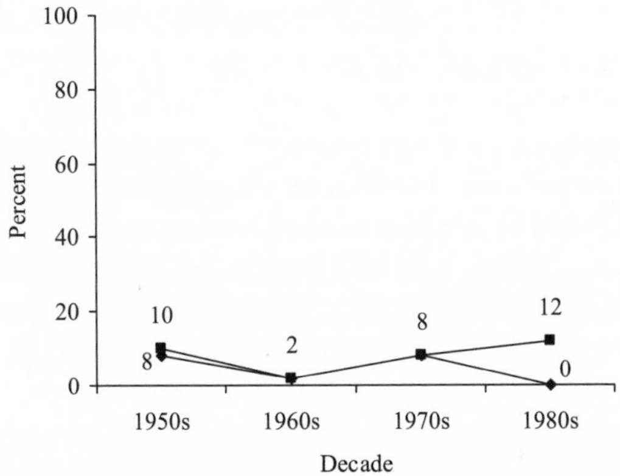

FIGURE 14 Exchanges for $10 or less as a percentage of total exchanges, Italian South Philadelphia and Toronto's Little Italy, 1940–1990. SOURCE: Philadelphia Deeds of Sale, Summary Sheets, 1950–1990, Office of Records, City Hall, Philadelphia; Toronto Deeds of Sale, 1940–1990, Office of Land Title/Land Registry, Ontario Ministry of Consumer Affairs, Toronto. *N* = 889 (Philadelphia 358, Toronto 531).

ously illuminates a troubled housing market and the creative responses of area residents. Neighborhoods similar to Italian South Philadelphia were especially hard hit by economic downturn and racial tension in postwar American cities. Often lacking the restrictive covenants and zoning regulations that barred African Americans from many suburban areas and located in proximity to growing African American populations, working-class Catholic neighborhoods became the site of intense, and often violent, racial tension in the decades after World War II.[35] In

Philadelphia, the Commission on Human Relations recorded hundreds of instances of discrimination and violence in the housing market in the 1960s and 1970s. In a summary of its research in the mid-1960s, the commission reported problems of racial exclusion in West, North, Northeast, and Central Philadelphia.[36] South Philadelphia was the kind of area where similar trouble might have been expected. Indeed, as noted earlier, racial tension bubbled forth in various ways in South Philadelphia. And yet, when it came to the housing market, the commission found little evidence of discrimination or open hostility.[37]

Local practices such as housing gifts helped to prevent violence in a housing market where racial turnover was a constant concern. Some homeowners in Annunciation and St. Thomas elected to give their properties away rather than sell outside the social networks of their struggling neighborhood. Such decisions were encouraged by local pressure. In St. Thomas Parish, on the outer edge of Italian residential dominance, attempts to prevent racial turnover met with mixed results. Ida G. recalls: "The woman across the street, she was the first one that broke the block. She sold it to the black." Concerned neighbors attempted to intervene, but, in this case, too late: "They asked her, 'why did you sell it to the black?' She did have white buyer, but she got more money from the black."[38] Residents also monitored property use in their neighborhoods. Anthony S. and his neighbors endeavored to prevent the conversion of single-family homes into rental units, which they feared might attract African American tenants: "As soon as somebody is fixing up a house we got two or three guys that go over and say, 'You can't have apartments up this street.'" According to Anthony, such tactics preserved the neighborhood in the 1960s and 1970s: "All Italians were coming in, that's all we had up here."[39] In Italian South Philadelphia, owners who did not conform to community standards had to explain themselves to disapproving neighbors. In this context, gifting a house was a viable alternative for many South Philadelphians.

Together, these practices were highly effective in maintaining racial boundaries in South Philadelphia. In 1980, African Americans constituted just 1 percent of the population of Annunciation Parish. Exclusion in St. Thomas Parish was less complete: in the parish as a whole, 66 percent of residents in 1980 were identified by the census as African American. However, comparison of the eastern and western halves of St. Thomas Parish suggests the enduring strength of exclusionary practices. In 1980, the western half of the parish, never heavily populated by Italians, was 95 percent African American. In the eastern half of the parish, where 43 percent of residents reported Italian ancestry, 42 percent of the population was African America.[40] On a smaller geographic scale, street fronts such as Ida G.'s remained predominantly Italian, with only one or two African American households.

Thus, local Italian social networks facilitated a relatively subtle form of racial exclusion. Indeed, some South Philadelphians came to think of their real estate practices as only accidentally exclusionary. Cynthia C. reflected that houses remained in Italian hands:

> Not for any reason, but because that was the way it was happening . . . People wanted to stay around their family and their friends . . . It could very well have been [that] I was an Italian selling to an Italian, but not because I didn't want to sell to anyone else.[41]

Racial exclusion was subtle because it overlapped with other priorities.

Keeping houses within Italian networks, South Philadelphians expressed the dominance of social expectations in a market in which little profit could be anticipated. Rather than buying and selling homes for profit, Italian South Philadelphians valued houses for their social use; by giving properties to their children, Italian South Philadelphians helped make the area a place where they could dwell among friends and family for their remaining days. With the suburbs drawing many of their children away, aging South Philadelphians likely saw the gift of a property as a way to keep their children nearby. For those who were leaving the area entirely, selling a house within the area's social networks was a final act of loyalty to the old neighborhood. With neighbors observing the market and patrolling racial boundaries, vendors saw more to lose in transgressing social norms than they had to gain in selling to a wider market.[42]

In Toronto's Little Italy, homeowners operated with very different expectations. Properties were seldom given as gifts, and Italians showed no special propensity to give their properties away.[43] Even when the seller and buyer carried the same surname, only 36 percent of Little Italy's properties sold for less than $10. People selling to family members did provide housing at a discount—excluding gifts, the prices of houses sold between family members averaged 54 percent of those sold between people with different names—but most could not afford to part with the valuable houses of Little Italy free of charge. Householders in Little Italy had acquired their homes at considerable expense and could expect substantial profit from their resale. Surely Italian Torontonians shared with their Philadelphia peers a desire to be surrounded by friends and family, but they did not realize this aim through mechanisms of neighborhood dominance.

Despite the significant differences between the two Italian enclaves, Italians in Toronto's Little Italy, like their counterparts in South Philadelphia, sold disproportionately to fellow Italians. They did so in part because of the influence of real estate agents and lenders in many housing transactions. These figures, much like buyers and sellers, are part of the social history of the real estate market. Agents brought buyers and sellers together while lenders made transactions possible. Patterns of exchange cannot be fully understood without the inclusion of these additional parties. As in the rest of the real estate market, agents and lenders played distinctive roles in each context and had a different role in connecting ethnicity with housing exchange. The agents and lenders in South Philadelphia fit the larger pattern of defensive localism in the area; agents and lenders in Toronto helped to ensure that ethnicity continued to matter in a market where the preservation of local turf was scarcely a concern.

Real estate agents in Toronto's Little Italy often benefited from sharing ethnic origins with their clients. John D. recalls the importance of ethnicity to his early ventures into the real estate market in the 1970s. Because of his recent arrival from Italy, his employer, a non-Italian firm, assigned John to the Italian corridor running from College Street to St. Clair Avenue. His language skills proved an important asset. In the early 1970s, John drummed up much of his business by canvassing door-to-door in search of clients: "Although I had clients of other nationalities . . . [it] was a lot more comfortable for me to speak in Italian because I could relate more, and the people themselves, they felt that they could open their mind and tell me things." The connection ran still deeper when he encountered people from his home region in Italy: "You see if I go to somebody and a coincidence is Abruzzese . . . I can make jokes about Abruzzese culture." Since his work required intimacy with his clients—"even have a meal with them or a glass of wine"—such connections proved vital.[44]

Real estate agents such as John served a crucial function in facilitating housing exchanges in Toronto's Little Italy. In addition to listing and showing properties, they often arranged the "vendor take-back" mortgages that assumed dominance in the area. These mortgages, wherein the buyer obtained a loan from the seller, were essential to real estate exchange in Toronto's Little Italy. Between 1940 and 1990, banking or lending institutions tendered only 18 percent of sampled mortgages in Toronto's Little Italy as compared to 87 percent in Italian South Philadelphia.[45] The contrast is even sharper before the 1980s, when lending institutions held only 8 percent of the total loan market in Toronto's Little Italy. The scarcity of formal institutional loans in Little Italy during this period is partially attributable to government policy and industry decisions that favored loans on new construction and left areas of the center city deprived of capital.[46] At the same time, many recent Italian immigrants, who often arrived with little equity or credit history, were particularly poor candidates for institutional loans. For the most part, sellers filled the void left by formal institutions. Home sellers in Little Italy provided three-quarters of the first mortgages in the area. The mortgages issued by sellers were by no means insignificant: although they were typically short-term—requiring payment in full after an average of six years—they carried a mean value of over $45,000 (USD 1989) and comprised almost half of the total amount exchanged for the properties.[47]

Agents specializing in work with Italian immigrants became experts in arranging vendor take-back mortgages, much to the benefit of sellers, buyers, and themselves. Most often, vendor take-back mortgages were subsequently sold to a mortgage broker. The brokers charged a premium, paying, for example, only 16,500 for an 18,000 mortgage, but the vendor was rewarded with cash that could immediately be put to use in purchasing another property.[48] Because of such mortgages, vendors widened their potential market, especially among the immigrants moving into the area. In the case of a default they remained responsible for repossessing the property, but with values steadily rising this seldom seemed a significant encumbrance. Italian buyers, meanwhile, benefited from credit that would otherwise have been unattainable. Partially because of the complexity of such arrangements, agents such as John D., who had close ties to reliable brokers, cemented their roles as key conduits to the exchange of property in Italian Toronto.

Real estate professionals did not restrict their business to the confines of Little Italy. John D. lived outside of Little Italy when he began to work there, and his firm, which was located near St. Clair, gave him responsibility for a large section of the city. Other agents active in the area were also broadly based. In the 1970s and 1980s, large Italian firms—such as Nardi and Racanelli Real Estate—sold housing in Italian districts throughout the city, commanding large ads in the *Corriere Canadese*. Even larger firms—such as A. E. Lepage, Canada Trust, and Montreal Trust—advertised in Italian-language press and placed ads for houses in Little Italy in the city's widest circulating paper, the *Toronto Star*.[49] In 1990, Royal LePage, which listed a number of St. Francis properties in the *Star,* boasted 370 branch offices in North America and over 10,000 employees.[50]

Large, non-Italian firms often took into account the ethnic makeup of their clientele. Guaranteed Trust listed its St. Francis properties with an agent named Giammattolo and Montreal Trust with Nathan Flavia, while LePage's Manuel Painco may have had connections with the area's growing Portuguese population.[51] However, in Toronto, ethnic ties in the real estate market did not coincide with local bonds. Realtors such as Nathan Flavia and John D. participated in real estate networks stretching across the city. Ethnic, but not local, bonds between realtors and the areas where they operated fit the broader social structure of Little Italy's real estate market. Rather than facilitating local exclusion, ethnic bonds stretched across the city linking Italian Canadians in profitable real estate transactions.

When they sought the help of professionals, Italian South Philadelphians preferred local businesses.[52] An impression of the limits of professional real estate in Italian South Philadelphia can be gleaned from the real estate section of the city's widest circulating paper—the Sunday edition of the *Philadelphia Inquirer.*[53] Many properties, of course, would not have been listed in the city's largest paper. Italian South Philadelphians, I have argued, seldom aimed for the widest field of potential buyers. Nonetheless, a search of the first *Inquirer* real estate section in every month in 1950, 1960, 1970, 1980, and 1990 yielded a total of 217 listings in Annunciation and St. Thomas. This sample is deliberately unrepresentative. The agents listing with the widest audience in the city were the least likely to be rooted exclusively in South Philadelphia.

If indeed the agents in the *Inquirer* were the most wide ranging in the area, then the professionals in Italian South Philadelphia were every bit as local, and almost as Italian, as their clients. In the early postwar era, the most cited South Philadelphia agent in the *Inquirer* was Burton C. Simon, whose building and loan association also stood among the top lenders. Although Simon did not have an Italian surname, in 1950 the agency was located at Twentieth and Passyunk, one of South Philadelphia's main commercial corridors. Tarsatana and Izzi, on South Eighteenth Street, was the second-most-cited agency.[54] Twenty years later, an Italian American firm, Grasso-Tori Realty Company, with three locations in South Philadelphia, including one at Seventeenth and Passyunk, had achieved unprecedented dominance. In 1970, the agency claimed one in two properties that named an agent.[55] In all, South Philadelphia firms represented more than 80 percent of listings in Annunciation and St. Thomas between 1950 and 1970, and just over half of the properties listed with firms bearing Italian names. The outer limits of real estate networks ended very close to home.

Local firms showed little sign of relinquishing clients when the market improved in the 1980s. In 1980 and 1990, most firms listing homes in the two Italian parishes were located in South Philadelphia, and Italian firms, such as A. R. Fanelli Realty, remained prominent. However, the data from these dates are unreliable, since few properties were listed and fewer still indicated their realtor. Indeed, the paucity of advertisements in 1980 and 1990 offered perhaps the strongest indication of lingering localism. In 1950, 1960, and 1970, sampled papers ran an average of sixty-one advertisements per year, but in 1980 and 1990 the number slipped to seventeen. As the market for houses heated up, South Philadelphians ceased to advertise in the largest city newspaper's real estate section.

Firms within the geographic boundaries of South Philadelphia also lay within social boundaries. Cynthia C. recalled "your neighborhoods included the businesses. If a person lived in a certain area . . . they would want to go to a realtor in their neighborhood because they felt they knew the neighborhood and they could trust them."[56] Local networks provided clients but also made agents accountable to other neighborhood residents. Noel Smorto's testimony to the Commission on Human Relations revealed that residents moved decisively to prevent local realtors from breaking social norms—a realtor who threatened integration faced hostility at work and at home. The owners of Grasso-Tori Realty also indicated that failure to preserve local racial boundaries resulted in violent threats.[57] By pressuring real estate agents, local residents who were not actually selling houses could gain influence over exchanges in South Philadelphia.

Because they answered to residents of the area, local realtors were crucial allies in the efforts of Italian South Philadelphians to preserve their racially exclusive neighborhood. Although Smorto described himself as a victim of local social pressure, he also stood among the guardians of the area. Purporting that African Americans made neighborhoods uninhabitable for other city residents, Smorto contended, "They claim we are depriving them of the right to live where they want to live. We are saying they are doing it to us."[58] In keeping with such views, real estate agents safeguarded Italian South Philadelphia in the daily operations of their businesses. Both Smorto and agents at Grasso-Tori Realty steered African American clients to the western section of South Philadelphia, already dominated by black residents; to the east, houses were reserved for Italian Americans.[59]

Lending institutions in South Philadelphia also fit within the social ties of the Italian neighborhood. Borrowers with Italian surnames tended to use local savings and loans, often with Italian South Philadelphia connections. Of the top ten lenders to Italian buyers in Annunciation and St. Thomas Parishes, nine were savings and loan associations, including the Marconi Savings and Loan and the Savings and Loan of South Philadelphia; another top lender was Sons of Italy Bank and Trust. Of the top ten, half had offices in other parts of the city; the rest were exclusively South Philadelphia institutions with head offices, for the most part, on South Broad Street. These borrowing patterns set Italians apart from the non-Italian minority in the area. All but two of the top ten lenders to non-Italians in Annunciation and St. Thomas had a center-city branch, and none featured branches only in South Philadelphia. Only one savings association appeared in the top ten for non-Italians, and only one institution was in the top ten for both Italians and non-Italians.[60] People of Italian extraction exchanged real estate in a peculiarly Italian world. Buyers and sellers were local and Italian. So too were the professionals they entrusted to work in the area. Ethnic social networks both facilitated and constrained exchanges of real estate in South Philadelphia.

Housing markets were integral to the responses of ethnic neighborhoods to urban change. Postwar economic decline and the suburbanization of the city's white population froze property values in South Philadelphia. In response, local residents valorized Italian neighborhood and mobilized ethnic bonds. Guarding real estate within networks of family and ethnicity enabled parishioners to insulate their neighborhood against racial change with few direct clashes with African Americans. In a city dividing along racial and class lines, ethnicity became a way of delineating territory. Although such tactics were designed in part to protect properties from the devaluation imagined to accompany African American settlement, racial territorialism in Philadelphia likely contributed to the stagnation of the residential real estate market. Yet, Italian South Philadelphians might have had few better options. Integrated neighborhoods were exceedingly rare in postwar Philadelphia, and, between World War II and the 1980s, no integrated neighborhood experienced a real estate boom. The wider context of economic decline and racial division doomed the local real estate market. Although Italian South Philadelphians participated in creating these patterns, they could not have easily reversed them. Instead, Italian South Philadelphians molded their expectations to fit their circumstances. Rather than regarding property as a source of economic profit, they emphasized the social benefits of Italian neighborhood life.

In Toronto's Little Italy, rising property values gave concrete expression to urban prosperity, but the scarcity of lending institutions in older urban neighborhoods encouraged Italians to look to each other for capital. For Italian newcomers in Toronto, housing exchanges were part of a larger economic strategy that included immigration itself. For many, prosperity was far from assured, and housing became a key economic resource. Just as they had in the process of migration and settlement, networks among coethnics proved crucial to housing exchanges in Toronto. Although they were not defending urban turf, Italian Torontonians made use of ethnicity as they navigated the real estate market.

Housing sales in Italian South Philadelphia and Toronto's Little Italy illustrate the variable interaction of social experience and economic exchange. Housing markets shaped Italian ethnic life in a twofold fashion, imposing constraints while also opening avenues of response. Neighborhood residents had little say in the citywide trends that often determined their housing values. But Italians homeowners made choices when it came to selling and purchasing property. The decision of residents in each locale to exchange real estate within disproportionately Italian markets reflected the strategic use of ethnicity. Italians chose this response to urban economic shifts, and the decisions that they made in the real estate market carried powerful implications for the rest of social life.

3 Invitations and Boundaries: Patterns of Religious Participation

The Catholic Church stood at the center of postwar social life in Toronto's Little Italy and Italian South Philadelphia. In the decades after World War II, thousands of Italians crowded into churches every Sunday, and hundreds more came throughout the week for a wide variety of activities. As key institutional nodes, churches drew people of Italian origins together, helping to preserve the relevance of ethnicity in social life. Yet, as with the housing market, Italian religious practices in Philadelphia and Toronto were by no means identical. The associational meetings, hockey league matches, and street processions of St. Agnes/St. Francis in Toronto's Little Italy played host to a geographically elastic Italian ethnicity. By contrast, in South Philadelphia's Annunciation and St. Thomas, religious life reinforced parish boundaries.[1] The history of Catholic traditions in Little Italy and Italian South Philadelphia sheds light on the institutional dimensions of postwar ethnic community. The Catholic Church, though a common ballast of social life in both enclaves, played a crucial role in differentiating Italian South Philadelphia from Toronto's Little Italy.

Historians of North American cities have understood Catholic parishes as territorial institutions. The boundaries drawn on Archdiocesan maps lend credence to the view that parishes separate sections of the city from one another, dividing urban space into distinct segments. Even North America's national parishes—defined primarily by the ethnic populations they serve—have geographic boundaries on the assumption that most parishioners live within a neighborhood surrounding the church. Catholic parishes are fundamentally immobile; indeed, their religious significance is tied up with geographic permanence. Whereas Protestants and Jews could relocate churches, synagogues, and Sunday schools to serve their migrating memberships, Catholic parishes were rooted in place.[2]

A comparison of St. Agnes/St. Francis in Toronto's Little Italy and Annunciation and St. Thomas in Italian South Philadelphia complicates this picture. Although both Roman Catholic archdioceses marked Italian parish boundaries on their respective maps, the lines had different meanings in the two contexts. In Italian South Philadelphia parish boundaries demarcated the territorial limits seen elsewhere in studies of American Catholics, but the parish lines of Toronto's Little Italy designated an Italian Canadian space shared by parishioners and outsiders. In Italian South Philadelphia, parish boundaries set people and places apart from one another. In Toronto's Little Italy, parish lines created a space shared by the dispersed Italian population of the city. The relationship between spaces and people on either side of parish boundaries, therefore, differed in accordance with the local context of Catholic life; the spatial meaning of the parish reflected the wide array of urban social interactions of which it was a part.

Ethnicity and religion were inextricably interwoven in Toronto's Little Italy and Italian South Philadelphia. Catholic churches, and their many associations and clubs, shaped daily life in the two neighborhoods for decades after World War II. As the churches left different urban footprints, so too did Italian ethnicity. This chapter first explores the role of the church as a site of social connection in both Italian South Philadelphia and Toronto's Little Italy and then addresses the geographic reach of these connections.

Bustle and solemnity met at the church doors in postwar Little Italy and Italian South Philadelphia, as thousands of parishioners crowded into some ten masses every Sunday. In Toronto's Little Italy, the crowds surprised and overwhelmed the priests. In 1958, Father Vincent Mele at St. Agnes wrote the archbishop to describe the overcrowding: "Half of the people are standing in the aisles, in the side of the chapel of St. Anthony, in the back of the church and in the choir. This cannot continue."[3] Yet, with Italian immigrants pouring into Toronto, the problem did continue. Four years later, Father Adoph Baldolini wrote the archdiocese echoing his predecessor's concerns: "There are still one hundred to one hundred and fifty children standing at each Mass."[4] On Christmas in 1964, St. Agnes's priests remarked that "with the church packed like sardines for the Midnight Mass, many fainted during the services."[5] In response to such concerns, the archdiocese granted Italian parishioners the use of the larger St. Francis, a church standing a block north of St. Agnes.[6] Participation remained strong thereafter. In the 1970s and 1980s priests reported weekly church attendance in the thousands, numbers well exceeding 50 percent of the parishioners.[7]

In South Philadelphia, church attendance was equally strong, though it was the baby boom, rather than immigration, that increased the number of parishioners. Annunciation churchgoers recall that in the early postwar years, "Sunday Mass was so crowded you had to get there early to get a seat."[8] At St. Thomas, parishioners contributed twenty-four cents to assure themselves of seating, and priests "put folding chairs along the pews" to accommodate the crowds.[9] High participation rates in Philadelphia also persisted until the 1980s. At the beginning of that decade, priests in the two parishes estimated that more than 45 percent of parishioners attended on a weekly basis.[10]

If Sundays called more parishioners to church than any other day, the institution was nonetheless a center of activity for the rest of the week. Parishioners in both cities assembled on parish streets for feasts and festivals of religious and cultural significance and met for church bowling and hockey leagues, card and bingo nights, dances, and cultural festivals, as well as for a wide array of social clubs that made the parish a setting for Italian community in ways not always tied directly to religion or limited to the devout.[11]

In Toronto's Little Italy, the importance of the church found vivid expression in 1970, as popular discontent erupted over the replacement of Father Ambrose De Luca, the St. Francis priest. Although the rotation of priests was a routine administrative convention of the Franciscan order based in New York, parishioners "descended upon" Toronto auxiliary bishop Thomas Fulton with a letter describing "the best spiritual leader we have ever known," along with a petition, some eighty-five pages and four thousand signatures long, demanding De Luca's return. One parishioner, writing directly to Franciscan superiors, warned that her peers had only barely been restrained from more radical action: "We had to stop the people from going to the Bishop. They were going to strike, as they called it."[12] Their request was not granted, but Fulton was impressed by the enthusiasm of the parishioners for their former priest: "While I was aware of his dynamic personality, I had no idea of the magnitude of his popularity."[13]

De Luca's popularity may indeed have reflected his personal charisma, but it also underlined the wider importance of the church to residents of Little Italy. Well before De Luca's arrival, the church already played a pivotal role in social life. A letter from an Italian Canadian parishioner to the archbishop in 1935 described the range of roles played by clergy in Little Italy at that time:

> They built for our young people a hall were [*sic*] they could have their dances [and] socials or any kind of . . . party under the supervision of the good priests and it was a great success . . . The young people could meet one other [*sic*] and there [*sic*] parents would know were [*sic*] to find them. They also built under the church club rooms for the men . . . what a blessing these club rooms were. This was a good way the Dear Salesian Fathers had of reaching the men hearts [*sic*]. They would go down to the club rooms and mingle with the men dropping a good word here and there, and they surely sowed the good seed.[14]

Whatever the success of religious lessons sprinkled into card games, the diversity of events at the church made it central to social life. The author of the letter, concerned that such activities would cease with the transfer of the church from Salesian to Franciscan care, need not have worried. The pattern of institutional diversity and vitality of these early years was preserved into the postwar era. Longtime priest Father Riccardo Politicchia saw social clubs as a central function of the church in the almost thirty-five years, between 1934 and 1969, that he spent at St. Agnes/St. Francis.[15] Politicchia, who by his death in 1980 had attained iconic status within Toronto's Italian community, had a hand in the creation of dozens of social clubs, including the Columbus Boys' Club, the Young People's Club, the Italian Hockey League, the St. Agnes Boxing Club, the Catholic Action Congress, the Catholic Action Group, the St. Agnes Society, the Circolo Calabrese, the Catholic Women's League, the St. Agnes Bowling League, and the St. Agnes Recreation Club (fig. 15).[16]

Activities in and surrounding the church brought Italian Canadians together, forging connections that served a variety of purposes. An Italian immigrant who arrived in 1948 later recalled that as the director of a church choir he made contacts that enabled him to escape working as a furniture salesman and return to teaching, which had been his profession in his hometown near Treviso. Another used the church as a launching point for an association for the large number of immigrants from Pisticci. In the early 1940s, the Loggia Pisticci "had no place where

FIGURE 15 Priest drops the puck to begin a Catholic league hockey game, Toronto, circa 1950s. SOURCE: "Silver Jubilee Commemoration at St. Agnes Church, Grace and Dundas Streets Toronto, Canada," Parish History: Franciscan's Twenty-fifth Anniversary Publication, 1959, St. Agnes Parish Papers. Courtesy Archives of the R. C. Archdiocese of Toronto (ARCAT).

to start," so he and his *paesani* turned to St. Agnes Church, where "they had a little hall . . . I think we used to pay one dollar . . . and that's how we started." Many Italian Canadians met their spouses at church social functions. Summing up the period from the late 1940s to the late 1970s, one immigrant reflected, "I don't recall in the thirty years that I am in Canada ever having gone through a period without having been part of some club or another."[17]

While no official statistics were collected on this kind of informal church participation, records of the events confirm significant popular involvement in secular church activities. St. Francis ran Saturday night bingo games at the church in the 1960s that Father Nincheri described to the archbishop as necessary "extra income" for running the parish. The economics of bingo night—which entitled players to some twenty games for the one dollar price of admission and awarded more than $1,000 in cash prizes—meant that it required substantial attendance to turn a profit. If, as these figures suggest, thousands of bingo players assembled every Saturday night at St. Francis, then it is little wonder that Nincheri felt compelled to assure the archdiocese that "the St. Francis bingo is conducted in a very honest manner and we've never had any trouble whatsoever with the police department."[18] In 1984, the St. Francis Bazaar included a raffle with similarly high levels of participation. The raffle promised a first place prize of two plane tickets to Rome and $1,000 spending money as well as four prizes of $100.[19] Even if the plane tickets were donated rather than purchased, the raffle required $1,400 in ticket sales to break even; participation was likely in the hundreds, if not thousands. Mass Italian participation in church events expressed the religious and social significance of the parish in daily life.

In Italian South Philadelphia, the church claimed a similar place in local social life. St. Thomas and Annunciation Parishes offered a seemingly endless array of social events for parishioners. Weekly activities included bingo nights, bowling and baseball, lectures on religious themes, meetings of the women's Sodality of the Blessed Virgin Mary, and the men's Holy Name Society, youth and seniors' club events, dances, and guitar lessons. More notable events punctuated the monthly calendar, including local pilgrimages and bus trips, drama group productions and variety shows, society breakfasts and dinners, picnics, retreats, penny auctions, and varied religious feasts.[20] Attendance at such events was likely high: in 1950 the nine largest associations at the two churches claimed over 5,000 members; twenty years later participation in the adult associations had slipped significantly, especially at St. Thomas, but with the addition of the large Catholic Youth Organizations, clubs at the churches retained more than 4,700 members (fig. 16).[21]

As in Toronto, church documents suggest high levels of participation at social events in Italian South Philadelphia. Bingo nights made significant profits. Reflecting on the church's revenues from the 1950s to the 1970s, a nun at Annunciation Parish recalled that "bingo was one of the real biggies." Parishioners remember lining up, dinners in hand, "at 3 o'clock to play bingo at 7."[22] In the 1970s, the St. Thomas church bulletin announced that a bus would drive loops in the parish delivering parishioners to bingo and home again, and profits from such evenings routinely exceeded $1,000.[23] Carnivals to benefit Annunciation Parish gave away impressive prizes—cars, televisions, appliances, wristwatches, and cash—suggesting that the games of chance attracted thousands of parishioners from the 1950s to the 1980s.[24]

In all, the churches of Italian South Philadelphia and Toronto's Little Italy were similar in many ways, including in their vibrancy. Despite their similarities, however, the churches also exemplified the differences between the two Italian neighborhoods. Like housing exchanges, institutional life expressed the differences imposed by local urban environments. Italian South Philadelphia's social institutions buttressed

FIGURE 16 Nun supervises street play in South Philadelphia. SOURCE: *St. Martha's House/Houston House Community Center,* PC 36, circa 1968. Philadelphia, Temple University Archives, Urban Archives, Philadelphia, PA.

territoriality, while those in Toronto's Little Italy followed a different pattern.

A structural difference between the parishes in Canada and the United States encouraged different patterns of institutional life. Control of parochial schools provided American parishes with an additional pillar of neighborhood cohesion that was missing in Toronto, where Catholic schools were run publicly by the Separate School Board, a responsibility of the Metro government. Although previous generations of Italian South Philadelphians, like their coethnics elsewhere in America, had been slow to embrace formal education—both in parochial and public settings—by the postwar era parochial schools had become vital neighborhood institutions.[25] In Annunciation and St. Thomas virtually every church bulletin advertised dinners, performances, and raffles for the benefit of the parochial schools. The schools helped cement parish pride, as in 1948 when Annunciation parishioners were invited to celebrate the dedication of their parochial school: "Another historical milestone in the progress of the parish will be marked . . . with the official blessing of our new school."[26]

More than merely bolstering spirits, parish-run parochial schools reinforced neighborhood by bringing the children of the parish together under the supervision of their priest. In 1970, the priests in Annunciation and St. Thomas estimated that 80 percent of local children went to their parish grade school. Catholic secondary schools, attended by some 85 percent of high school-aged children from the two parishes, were outside the purview of the parish priests but performed a similar function.[27] South Philadelphia's St. John Neumann High School for Boys, just west of St. Thomas Parish, and Saint Maria Goretti High School for Girls, in the heart of Annunciation Parish, drew Italian South Philadelphia together, with students from the many parishes in the area.

In Toronto, the relationship between St. Agnes/St. Francis and the local parochial school was far more tenuous. In Canada, Catholic schooling was guaranteed by law and paid out of the public purse.[28] By the postwar era, parish priests held at best an ambiguous role in the administration of Toronto's large and growing Catholic school system.[29] Indeed, in 1963 the priest at St. Agnes, Adolfo Baldolini, complained that he felt "ill-informed and unwelcome in important school decisions," a predicament scarcely imaginable in the South Philadelphia parishes.[30] Moreover, relatively few students from the parish attended Catholic schools. Despite rising enrollment figures in the postwar era, only 27 percent of St. Francis children attended parochial elementary schools in 1970, and less than 2 percent went to Catholic high schools.[31] The parish at the center of Little Italy could place little faith in Catholic schools as a source of local cohesion.

In keeping with these structural arrangements, local cohesion was much less integral to St. Agnes/St. Francis than to the parishes in Italian South Philadelphia. The spatial contours of parish life are difficult to trace because the geographic span of church participation leaves little documentary record. The presence or absence of outsiders to the parish at social and religious events may have merited attention from parishioners and priests, but in most cases this attention left little evidence. Two intriguing exceptions to this rule facilitate comparison of the social geography of religious participation in Toronto's Little Italy and Italian South Philadelphia—street processions and parish publications.

Street processions served a range of social and religious purposes. Italians solemnly marching through the streets of Philadelphia and Toronto, like their counterparts in other centers of Italian settlement, expressed their faith and publicly evoked a wide range of concerns and aspirations. Participants in processions could simultaneously affirm and challenge structures of social and religious power.[32] Transplanted to North America, the street processions also expressed urban politics: parishioners walking city streets en masse could stake claims to territory and issue statements about inclusion and exclusion.[33] In Italian South Philadelphia and Toronto's Little Italy, street processions took very different forms, each expressive of the wider spatial dynamics of social life. Whereas the scale and scope of parish processions in Philadelphia reinforced territorialism, processions in Little Italy invited geographic outsiders, but ethnic insiders, into a shared celebratory space.

For the most part, processions in Italian South Philadelphia after World War II have been small-scale affairs, requiring little organization, leaving behind few documents, and garnering little attention, even from South Philadelphia's local press. The various parishes dotting Italian South Philadelphia each followed a distinct tradition, bringing forth their own statues on various saints' days and carrying them along short routes according to their own parish custom. In the spring, each parish held a May Day procession for its own schoolchildren that culminated when the veiled May Queen—an exemplary eighth grade girl selected by her peers or her teachers—crowned the statue of the Blessed Virgin Mary (fig. 17).[34] Annunciation parishioners recall that South Philadelphia tradition held that "everybody [each parish] does their own; they don't join with one another."[35] According to St. Thomas's priest and longtime parishioner, Father Taraborelli, South Philadelphia processions and parish carnivals have always been "more of an internal thing."[36] While devout Catholics from the parish often traveled to large processions outside the city—nationally renowned processions took place in Hammonton, New Jersey; Norristown, Pennsylvania; and Wilmington, Delaware—South Philadelphia parishes such as Annunciation and St. Thomas never hosted an affair of similar magnitude.[37] Instead, parish processions were made up of the residents of the parish; the meanings, and messages of such events were, for the most part, limited by parish boundaries.

In keeping with this broader sketch, Annunciation and St. Thomas have their own histories of parish processions and street activities. In Annunciation Parish, street processions were most common in the early postwar period, with the procession of the men's Holy Name Society for the Feast of Christ the King receiving the most attention in par-

FIGURE 17 May Day procession in Annunciation Parish, 1951. SOURCE: Personal photographs used courtesy of Cathy S.

ish publications. By the late 1960s, however, the procession had been dropped from the celebration of Christ the King, and no other procession subsequently received similar publicity. In St. Thomas, by contrast, processions gained momentum in the late 1970s with the arrival of Italian priests in the previously Irish-run church and the inauguration of an "Italian Festival" in the parish. The processions lasted until the mid-1980s, when the festival was disbanded. Closer attention to the processions in both parishes—and to the language parishioners and parish leaders used to describe them—tells a story of complicated spatial practices. Parish processions deepen and confirm a view of South Philadelphia parishes as territorial.

Annunciation parishioners held their procession for Christ the King in late autumn. Although members of the Holy Name Society, an association of devout men from the parish, occasionally joined their peers from the rest of Philadelphia for massive processions on the Benjamin Franklin Parkway, a thoroughfare on the northern edge of the city center and distant from South Philadelphia, their annual processions took place within the parish, and on a much smaller scale.[38] The parish procession for Christ the King invoked Catholic masculinity to proclaim the parish territory. The Holy Name Society was responsible for the religious and civic aspects of the celebration, which involved a special men's-only mass followed by a procession of the men through parish streets. Men of the parish were urged to join the procession to "manifest to our neighbors and to the world our love and loyalty to Christ the King."[39] In 1950, a procession of three hundred men through the parish was said to have offered "a demonstration of real manhood," and a year later, the parish calendar chided men who had not marched in the past: "Be a Real Man, show your love for Christ the King."[40]

Displays of Catholic manhood targeted a diverse audience. The march of the parish's most active men through the parish was surely designed to convince other men of the joys and status conferred by church participation. An all-male parade also affirmed the position of men as leaders in the public and private expression of Catholicism. This was a somewhat strained message, given the paucity of male participation in parish activities. Women's groups such as the sodality were more active and better enrolled than the Holy Name Society. In 1950, Annunciation Parish's sodality had 1,287 members to the Holy Name's 629.[41] In the church's battle to retain male participation and to offer its own vision for masculinity in the parishes, a public procession of devout men (with women and children encouraged to line the streets to watch their men pass by) was designed to send a powerful message.[42] The calendars emphasized the worldly battles of the Catholic man, battles to protect the parish against communists and secularists who threatened both the church and the nation. As the calendar reminded parishioners: "In these days when the forces against God are so active—yes, even in the United States—the success of these religious rallies is most important."[43]

An additional audience for Catholic manhood, which went unmentioned in parish publications, becomes evident once the routes of the processions are considered. In the early 1950s, processions spent a good deal of their time—more than one-third of the total procession route—tracing the territorial outlines of the parish (map 6).[44] The processions, therefore, were also a form of boundary marking.[45] As they marched along Federal, Reed, Broad, and Seventh Streets, the men of the parish were visible to those outside parish limits; what the calendar called an "outward demonstration" of love of Christ was also a demarcation of

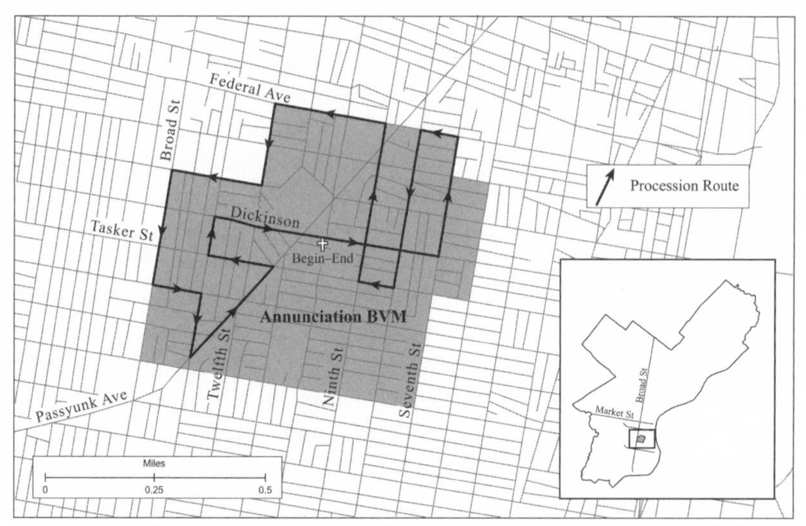

MAP 6 Christ the King Procession Route in Annunciation Parish, 1952. SOURCE: Annunciation Parish Calendar November 1952, Annunciation Parish Archives, Philadelphia.

urban turf.[46] In this context, emphasis on "manhood" takes on a different cast—the men had a gendered responsibility to proclaim parish boundaries to outsiders.[47]

Parish calendars offer little hint of the reasons for the waning of the Christ the King procession in the 1960s. The church's vision for Catholic manhood had long competed against other, more popular models of masculinity, as was suggested by the low levels of participation in the Holy Name Society. As these numbers dwindled further, the procession may have become unfeasible.[48] Yet, even after the procession disbanded, authors of the parish calendar continued to urge parishioners to use the day as inspiration for outward displays of their faith: "Christ showed us by His actions that His words were not empty and false. We too must show the world—our neighborhood—that our words are not empty and false. We must live our faith."[49] If the quieting of procession bands and the retirement of Holy Name banners deprived the day of its early postwar potency, Christ the King was still associated with an affirmation of local boundaries.

In contrast to its neighboring parish, St. Thomas's most notable procession of the postwar era, the Italian Festival procession, began in the 1970s and ran for a decade thereafter. As in Annunciation various small processions had characterized parish life for some time, but in July 1977 the parish calendar announced a procession on a grander scale. The authors of the calendar inquired: "Have you ever wondered what happened to the old-time Italian festivals that they used to have in South Philadelphia? The sound of Brass Bands walking through the streets, the taste of torrone candy, the smell of dough frying?"[50] St. Thomas Parish, the authors declared, would "re-create" such memories in its "first annual *Festa Italiana*" in October of that year. Accompanied by a carnival, a draw for $2,000, games of chance, a spaghetti dinner with two meatballs, as well as folk dances led by two Italian bands, the festival would allow parishioners to "work as a unit" and enjoy their heritage together, while simultaneously raising funds for the parish.[51] The parish calendar's announcement that Mayor Frank Rizzo, a champion of Italian South Philadelphia, would headline the distinguished guests surely raised expectations even further.[52]

The first festival was considered a tremendous success. A week later, the parish calendar exclaimed, "*Congratulations* . . . you showed that there is plenty of *life* and *spirit* in this *parish*. You joined together as a community in a celebration we are sure you will remember for a long time to come . . . it won't be our last." The profits secured in the event virtually assured that the festival would indeed be repeated. With all expenses paid, the parish had profited more than $17,000.[53]

The annual Italian festival continued for a decade in very similar form and with similar results. By all accounts the carnival was a parish affair—St. Thomas parishioners have little recollection of outsiders coming to the event, and members of other parishes report no knowledge of St. Thomas's Italian festival. And yet, the event remained remarkably successful. As late as 1986, the October procession, carnival, and entertainment netted the parish almost $17,000.[54] The parishioners, who numbered only 4,000 by 1990, must have participated at an astounding rate.[55] With other parish activities on the wane by the 1980s, the Italian festival remained a key source of revenue and pride.

As the original announcement had suggested, a street procession anchored the *Festa Italiana*. The parish calendar announced the route in advance, so that those along the way could decorate their houses and streets. The best remaining documents of the event are photographs taken in 1984, which reveal the mixed symbolism of the procession. Some portray the scenes expected from an Italian procession in an aging South Philadelphia parish. In one, middle-aged parishioners pull a saint, partially obscured by the dollar bills tacked by the devout to its yellow celebratory ribbons, past a white-capped brass band and the row houses of a typical South Philadelphia street (fig. 18, p. 74). In another, a figure of Jesus spreads its arms behind Mary in the back of a red pickup truck while parishioners affix ribbons to the statues in the car ahead (fig. 19, p. 74). A third shows the front of the procession: teenage girls carry the lead banner—"St. Thomas Aquinas"—while behind them older men shoulder the American and Italian flags and a banner for Saint Anthony (fig. 20, p. 75).

The parishioners carried their statues down the tiny streets surrounding the church—Chadwick, Bancroft, Sigel, and McClellan, streets hardly known to outsiders—and never touched the parish boundary line. Notably, the route remained entirely within the southeastern quadrant of the parish, within the area most densely inhabited by Italian parishioners—it never strayed west of Eighteenth Street, into the area with an African American majority.[56] Together the photographs and route tell a story of late twentieth-century parochialism—an Italian ethnic parish reclaiming street processions as a way to bring coethnics together in a celebration of their enduring neighborhood.

However, two further photographs give a quite different impression of the *Festa Italiana* procession. In one, a police patrol vehicle leads the way for an African American boy carrying the St. Thomas banner at the front of the procession (fig. 21, p. 75). In a second photograph, children pose in two long single-file lines in front of the church building. The student group is racially integrated; a significant minority appears to be African American. Three paper signs, held high on wooden poles, are visible in student hands. At the front of the group, a placard displaying the image of the American flag reads "Pray"; another sign shows the one time abolitionist symbol of the Liberty Bell beside the word "Love"; towards the end of the group, a sign calls for "Respect" (fig. 22, p. 75).[57] While the procession may have reminded some of "old time Italian festivals," the image more readily recalls "old-time" civil rights marches.

The photographs portray the children of St. Thomas Aquinas parochial school, key participants in the procession. The school had been integrated since the early 1970s, when the declining Italian population no longer sufficed to fill classrooms. The policy change stirred controversy in the parish, as some parishioners sought to remove their children from the integrated environment.[58] But the priests and parish publications proudly proclaimed a integrationist policy, boasting that "[o]ur school does not discriminate on the basis of race, color, national and ethnic origin."[59]

Efforts to include African Americans and other non-Italians in the activities of the parish expanded in the decade after the school integrated. In 1983, the calendar described a wider program of outreach to African American and Asian residents within parish boundaries, most of whom were not Catholic. "Due to the critical need in the community surrounding St. Thomas Parish," the calendar announced, "a program is being implemented to empower the people of the neighborhood." Here "community" and "neighborhood" included non-Catholics and nonparishioners, as the church requested volunteers to offer training in literacy skills.[60]

St. Thomas thus participated in a much wider movement in the postwar urban Northeast. As parishes lost their traditional Catholic bases and priests trained during and after the liberal turn of the Catholic Church in the 1960s took charge of parishes, churches often turned their attention to the new immigrants and underprivileged African Americans within their territorial boundaries.[61] A procession of schoolchildren through parish streets, signs demanding love and respect held high in both black and white hands, proclaimed the new direction of Catholic life, which pointed away from Italian insularity.

The *Festa Italiana* carried paradoxical meanings. The procession consciously recalled prewar urban Italian Catholic heritage and the practices that held Italian ethnicity together while proclaiming ownership of city territory. At the same time, the procession also expressed Catholic ambitions to provide a "beacon of hope" for the most marginalized residents of the city.[62] Territorialism was the common thread binding these divergent meanings together and sustaining the procession. The fissure between the liberal and reactionary tendencies of Catholicism in postwar American cities could be bridged by the procession winding through St. Thomas because all Catholics understood the parish as a territory.[63] If the meaning of territorialism varied—sometimes characterized by defense of boundaries and sometimes by the inclusion of all those within the parish sphere—they nonetheless shared a logic that set parts of the city apart from one another. Organizers and participants in the *Festa Italiana* procession understood St. Thomas as a neighborhood institution, one that served those within the lines drawn on Archdiocesan maps.

The marriage of such different conceptions of territory, however, was uneasy and unstable. According to Father Tarraborelli, who called an end to the processions in the early 1990s, "we couldn't go out on the streets because the neighborhood changed . . . I mean you'd have eighty-year-old women pushing statues around . . . they were getting harassed by the neighbors." In the same period, according to the priest, "black kids used to come and throw stones" at church buildings. Tarraborelli decided that a fundraising dinner could raise sufficient funds without the risk of exposure on the streets.[64] In this account, the ambiguous territorial meaning of the *Festa Italiana* is replaced with a simpler

FIGURES 18–22 St. Thomas *Festa Italiana* procession, 1983–1984. SOURCE: Personal photographs used courtesy of Father Taraborelli, St. Thomas Aquinas, Philadelphia, PA.

conception of turf: the shrinking Italian residential population lost control of parish streets. Italians could not hold processions on streets that they no longer possessed. This explanation of the demise of the *Festa* conspicuously ignores the ongoing presence of "black kids" within the school and the procession; in justifying his decision, the priest omits the ongoing contradictions of territorialism. In its inception, duration, and demise, the *Festa Italiana* exhibited the complex and shifting meanings of Catholic territorialism. At every stage, the *Festa* was defined and constrained by parish boundaries.

Whatever the ambiguities of territorialism in St. Thomas, the *Festa*

SAINT THOMAS
AQUINAS

SAINT THO
AQUIN
1719

PRAY

Italiana, like the processions of the Holy Name Society in Annunciation Parish, affirmed the importance of locality in social life. South Philadelphia was dotted with parishes dominated by Italian parishioners. St. Nicholas of Tolentine, St. Paul, St. Rita of Cascia, St. Philip Neri, and St. Edmund, among others, all served Italian South Philadelphians, but each did so separately. Parishioners recall no procession, church event, or celebration that regularly drew South Philadelphians across parish boundaries. Further, the thousands of residents of Italian origins who left South Philadelphia for the suburbs seldom saw the church as sufficient reason to return to the old neighborhood. Parishes in South Philadelphia primarily set, rather than bridged, boundaries.

Parishioners and pastors at St. Agnes/St. Francis in Toronto's Little Italy expected their institution to operate quite differently in city space. Rather than serving the neighborhood, St. Agnes/St. Francis was conceived as a geographically expansive institution. For special religious occasions and processions, the parish served as a gathering place for Italians spread across the city, Italians who otherwise attended their own neighborhood parishes. This function was especially visible in mid-June, when thousands arrived to venerate Saint Anthony, and by the final decades of the century, in early April, when Italian Canadians arrived in ever growing numbers for the Good Friday procession.

For Italian immigrants in post–World War II Toronto, Saint Anthony of Padua held special meaning. The story of the saint's shipwrecked arrival in Sicily resonated with thousands finding their ways from the docks of Halifax to Toronto's Little Italy.[65] One immigrant to postwar Toronto, Paul N., laughingly recalled the connection between Saint Anthony and his own voyage to Canada. When their family came to Canada, Paul's mother acquiesced to friends and family in Toronto who had implored her to bring Italian alcohol into the country. Having packed the alcohol in a large trunk, the family watched in horror as a customs official in Halifax proceeded to inspect their luggage. Just when he seemed poised to discover the booze, the family obtained an unexpected reprieve:

> [O]n the side of the trunk [containing the alcohol] my mother had put a picture of St. Anthony. So I guess during the voyage, the frame, the glass broke. So when the customs officer put his hand on that side of the trunk he cut his hand with the glass. So he put his hand out and he says, "Close that thing and get out of here!" So my mother said, "St. Anthony did me the miracle!"

FIGURE 23 First Communion class leads the St. Anthony Procession in St. Agnes, 1956. Source: "Silver Jubilee Commemoration at St. Agnes Church, Grace and Dundas Streets Toronto, Canada," Parish History: Franciscan's Twenty-fifth Anniversary Publication, 1959, St. Agnes Parish Papers. Courtesy Archives of the R.C. Archdiocese of Toronto (ARCAT).

Although Paul remembered these events with humor, he became one of thousands of Italians who annually attended the procession in honor of the saint. For Paul, processions expressed faith and recalled history. "We don't put it on to . . . have a parade, to have a show," he explains, "the people that participate . . . they really have that faith, they do it with faith, they do it with love."[66]

The city's leading Italian paper, the *Corriere Canadese,* regularly reported on the celebrations in Little Italy. Veneration of Saint Anthony began with a densely packed Mass, as parishioners and other faithful pressed together among elaborate floral installations to hear the pastors describe the miracles of Saint Anthony and extol the example he provided for "all emigrants." In the afternoon even larger numbers gathered outside the church for the procession. Thousands—including the members of social clubs carrying their handmade banners, musical bands playing Italian songs, and tiny brides and grooms marching neatly in single file to their First Communions—accompanied the statue of Saint Anthony through the streets of Little Italy to Bellwoods Park, where pastors delivered a panegyric to the saint, served Communion, and distributed the day's special ceremonial bread (figs. 23 and 24). Following the formal service, Saint Anthony's statue was returned to the church as bands played and the devoted celebrated late into the night.[67]

The *Corriere Canadese* regularly reported "dense" or "immense" crowds at the event. In 1958, the crowd was estimated at over 10,000 people, and three years later the paper noted that the "multitudes" had arrived from "every part of the city and its surroundings."[68] Although the devoted likely awaited no formal invitation, in 1967 the *Corriere*

FIGURE 24 A rainy St. Anthony Day in Toronto's Little Italy, 1969. Photograph courtesy of Vincenzo Pietropaolo.

Canadese published a letter inviting "Italians from all of Canada" to join the parishioners of St. Agnes in their veneration of the Saint.[69] The parish, rather than a place of exclusion, was a shared Italian Canadian space. The procession was a celebration of Italian life in Toronto marked by the connections between residents of Little Italy and others outside the parish. An article in the *Corriere* in 1975 took the part of an impressed visitor to the parish who, by chance encountering the lights, music, and crowds for St. Anthony, felt transported to "some Italian locality."[70]

Priests' accounts confirm that many at Saint Anthony's Day were guests. In 1965, the feast brought "a never-ending crowd of people from all over Toronto." According to the pastor, the devoted "kept [the clergy] busy far into the night with demands for blessings, candles and confessions." A year later, some 12,000 attended the masses and the procession, which were broadcast on television and radio, and in 1967, the pastors' log reported that the fifteen masses for the feast day were "filled to overflowing as all the devotees of St. Anthony came from every section of the city."[71] Enormous crowds for mass, confessions, and the processions imposed an extra burden on the pastorate. As the log noted in 1965, "Needless to say the fathers were dragging" after the feast.

Yet, little direct benefit could be expected from such events. In 1983, revenue from a lottery draw, year book advertising, donations, a bocce tournament, and drinks and games totaled $9,805.22. Set against expenses, the most significant of which were beer, porchetta, bands, and printing, the net profit was a scant $18.60.[72] But the disagreements likely to emerge when residents of one parish flood into another, imposing demands without offering support, never emerged. St. Agnes/ St. Francis was understood, by its parishioners, its pastors, and certainly by Italians in the rest of the metropolitan area, as a gathering place. By the 1980s, the Saint Anthony procession in Little Italy faced competition from other parishes: in 1977 eight other churches celebrated the Saint's day with a solemn walk through parish streets.[73] Yet, none of the others compared in size to the Saint Anthony procession in Little Italy. In 1986 the Little Italy procession drew some 30,000 people, while no other Saint Anthony procession mustered more than 2,000. Indeed, the procession for Saint Anthony at St. Francis Church was the second largest of the forty processions held for various saint's days and Catholic occasions in the city that year, and the next largest was less than a third of its size.[74]

The continued veneration of Saint Anthony at St. Francis Church attested to the ongoing connection between symbols of migration and the streets of Little Italy. When they gathered in the area every June with friends and family, Italians celebrated a place once crowded with the most recent arrivals to the city—a place that stood at the center of the migration experience. Saint Anthony, who offered his blessing to migrants in need, was also a symbol of the trials and triumphs of moving to Canada. As they walked the streets of Little Italy with Saint Anthony, Italians from across the metropolitan area venerated physical and metaphysical markers of their shared past.[75] As Paul N. put it: "Where do you have big celebrations? At your home . . . all the Italians that were here from before 1950, that's their home."[76]

However, even the Saint Anthony procession was dwarfed by the celebrations in honor of Good Friday, which were also held at St. Francis. The procession drew immense crowds throughout the 1970s, and by 1986 the observers were estimated at 100,000 (fig. 25). The procession itself included thirty associations with thousands of members, eight floats, six horses, and two musical bands.[77] A year later, the scene merited a colorful portrait from Toronto's Catholic Register. According to the Register, "mammoth crowds" of Italian Canadians gathered along College Street and the rest of the procession route for "street theatre" that turned "the narrow streets of Little Italy into a modern day

FIGURE 25 An immense crowd gathers for the Good Friday Procession, c. 1970. Photograph: *Toronto Star.*

Calvary."[78] The Good Friday procession recapitulated, in heightened form, the spatial dynamics of the Saint Anthony procession. The *Corriere Canadese* suggested that Good Friday processions of the mid-1980s included Italian ethnics from the entire metropolitan area, the rest of southern Ontario, and even the United States.[79] The processions brought people together as parish lines demarcated a shared ethnic space rather than geographic boundary.

For some, the scale of the Good Friday processions diluted their religious significance. "Now tell me," demands Francesca S., who comes from the eastern periphery of the city with her children and grandchildren to the processions in Little Italy, "what is spiritual about this big representation with a big parade of personalities in front?" Francesca believes that the audiences for the processions are "attracted by the costumes, by the crowd," rather than by religious devotion. Nonetheless, Francesca continues to travel to the parish for the processions. She has her "spiritual moment" at her own church, but then attends the procession in Little Italy for a "celebration of our roots."[80] Others, such as Paul N., who became involved in the organization of the procession, continue to see the event as religiously significant.[81] Vince P. who has photographed the processions since the late 1960s, views the processions as both religiously profound and socially significant: "[I]t's . . . an important religious moment . . . [and] it's a time of gathering."[82] Whatever their views on the religiosity of the Good Friday procession, Italians have continued to attend in remarkable numbers. Good Friday, like Saint Anthony's Day, has called a generation of Italians together across wide distances into a shared ethnic space at the center of the city.

Parishioners and priests seem at a loss to explain the emergence of Good Friday as the parish's preeminent procession. Perhaps Good Friday gained popularity due to its symbolic difference from the Saint Anthony procession. Rather than venerating an Italian saint and the experience of migration, the procession honored Christ, marking a day shared by all Catholics. The attention lavished upon the event by the Catholic Register suggests that Good Friday succeeds in drawing significant attention. Although they used a traditional Italian form—the street procession—the crowds gathering in Little Italy expressed a broadly Catholic message that may have appealed to the increasingly Canadian-born Italian community. Nonetheless, the participants in the Good Friday processions, as well as and the crowds that thronged into the parish to watch, remained overwhelmingly Italian in origins.

After 1970, Portuguese processions in the same area demonstrated that while St. Francis was a site of invitation for Italians, it was not exclusive Italian turf. The former Italian church, St. Agnes, closed its doors for three years in 1967 when Italians transferred to St. Francis, but then reopened as a Portuguese parish.[83] Shortly thereafter, the Portuguese parishioners organized their own processions through Little Italy. Although Portuguese processions never attained a size or stature similar to those of the Italian parish, they nonetheless exemplified the shared and heterogeneous social structure of the area.[84] Because Italian social connection moved fluidly through space—congregating in Little Italy for special events but then dispersing again through the city—the use of the same space by Portuguese residents in the veneration of their own saints posed no profound challenge to the place of Italian ethnicity in the area.

The processions in Little Italy and Italian South Philadelphia vividly expressed persistent aspects of community life. The meanings of parish boundaries to parishioners and outsiders were especially visible at the largest events at the churches—when thousands of Italians celebrated their shared religion and ethnicity en masse—but the processions reflected dynamics that operated well beyond these special occasions. Processions bespoke the ongoing connections among Italians. In Toronto, Italians a good distance from Little Italy remained connected to St. Agnes/St. Francis, even as Italian parishes were established elsewhere in the city. In Italian South Philadelphia, by contrast, religious participation was a local affair. Parish publications provide evidence that these spatial dynamics persisted in the rest of social life, even when statues of saints rested indoors and associational banners lay furled at meeting houses.

The advertisements included in parish publications are especially useful for mapping community life. Monthly calendars, festival books, and volumes for anniversaries and memorials were produced by Catholic parishes with the help of business people. Indeed, advertisements occupied more than half of the pages in many such volumes. The hundreds of names and addresses written into the ads give a sense of who could be counted on to support the church, providing a rare map of the geography of religious participation.[85]

The majority of advertisers in sampled parish publications in both cities bore Italian names. The figures were especially high in Philadelphia. With the exception of the 1937 volume, when Irish and German names lowered the Italian total to 62 percent, the vast majority of advertisers in the Philadelphia parish publications bore Italian names. In the 1962 monthly calendar of Annunciation Parish, the figure reached a high of 95 percent; the volume at St. Thomas two years earlier was not far behind, at 90 percent. In Toronto's Little Italy the ads showed somewhat more ethnic diversity: 66 percent of advertisers in 1959, and 77 percent in 1978 bore Italian surnames.[86]

Thus, Italian-owned businesses offered the churches financial support in both locales. Italian business owners likely advertised out of their own connection with the churches. Some may have seen their ads as donations. Indeed, the volumes listed many donations by private individuals. However, it seems that the advertisements were often more than this: repeated reminders in church publications urged parishioners to frequent the stores that supported their parish. The pastors, at least, seemed to take the advertisements seriously. Elaborate and specific advertisements suggest that businesses also expected some return on their investments. In contexts where ethnicity guided economic and commercial relations they, more than non-Italian businesspeople, could expect to earn customers through the parish volumes.

While advertisers in Toronto were somewhat less likely to have Italian names, they were more likely to target a specifically Italian clientele. In 1959, this tendency was especially pronounced, as one in five advertised in Italian. Many among these offered services for recent immigrants. Thomas Studios, for example, advertised "Foto per passporto," to those who might have been considering a visit home, as well as "Foto di bambini," for those who might choose instead to send a picture to relatives in the old country. Cianfarani Travel Agency advertised translations and money transfers and offered loans for prepaid tickets that might be used to bring relatives to Canada. By 1978, when Italians names were a more dominant presence among the ads, only 6 percent were written in Italian.[87] Fewer businesses offered goods and services specifically targeting new immigrants, but many still advertised their ethnic ties. Purveyors of food, such as Unico—whose advertisement assured, "When it comes to food, we speak your language"—were especially explicit in their evocations of Italian roots, but most vendors merely named Italian proprietors and workers to publicize their membership in Italian community.

In Italian South Philadelphia, advertisers were more likely to name Italian owners but less likely to run ads that otherwise signaled Italian ethnicity. Among the hundreds of advertisements in the four Philadelphia volumes, only one, for Marra's Restaurant, used Italian to indicate that it was a "vera pizzeria Napoletana," and even Marra's ad quickly added, in English, "Raviola and spaghetti our specialty." A smattering of advertisers indicated that they offered Italian food, especially baked goods, and one business, La Cantina Ristorante promised "Authentic Northern Italian Cuisine with a Fine Continental Flair." However, for the most part, the name of the proprietor sufficed to indicate Italian ethnicity.

Although ethnic networks played a similar role in sustaining church activities in both cities, the geographic span of their networks differed. In Philadelphia, the businesses supporting Annunciation and St. Thomas were located in the shadow of their steeples. In Toronto, advertisers in parish volumes were dispersed throughout the city and beyond. In Philadelphia, the shortest distances between businesses and the churches came in the 1960s, when both Annunciation and St. Thomas featured the ads of businesses that were located, on average, less than half a mile from the church buildings (fig. 26).[88] In St. Thomas's large Jubilee volume, five advertisers were located within a block of the church: to the north on Seventeenth Street, Cora-Lee Millinery Shoppe sold costume jewelry, hosiery, gloves and handbags; in a cluster to the west on Eighteenth Street, Joseph DiMaggio advertised the latest in household appliances, Nicholas L. Bertolino ran a pharmacy, Ralph's Gift Shop offered novelties, toys, glassware, and clothing, and Charlie & Tony's Fruit and Produce Market promised fresh fish on Fridays. The advertisers in

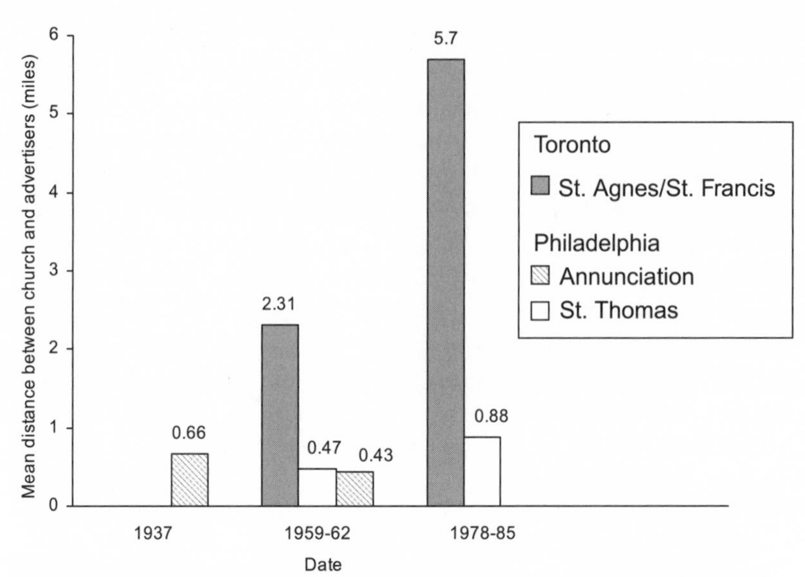

FIGURE 26 Distance between churches and the businesses that advertised in their publications, Italian South Philadelphia and Toronto's Little Italy, selected dates. Source: "The Annunciation Church, Philadelphia Pa., 1860–1937," box 4, group 102, Parish Histories, PAHRC; "Saint Thomas Aquinas Parish Diamond Jubilee," 1960, box 48, group 102, Parish Histories, PAHRC; Annunciation Monthly Calendar, January 13, 1962, Annunciation Parish Archives, Philadelphia; "St. Thomas Aquinas Parish, 1885–1985," box 48, group 102, Parish Histories, PAHRC; "Silver Jubilee Commemoration at Saint Agnes Church, Grace and Dundas Streets Toronto, Canada," Parish History: Franciscan's Twenty-fifth Anniversary Publication, 1959, St. Agnes Parish Papers, ARCAT; "Diamond Jubilee Celebration, St. Francis/Old St. Agnes Parish Community, 1903–1978," Publications: Seventy-fifth Anniversary Book, St. Francis Parish Papers, ARCAT. *N* = 465 (Philadelphia 301; Toronto 164).

the volume spiraled outward from the church, with sixty-seven of them, just a shade under half, within the parish limits (map 7). An additional thirteen were located in neighboring Annunciation Parish; almost all, 91 percent, were located in the southern section of Italian South Philadelphia, between Washington and Oregon Avenues. Two years later, Annunciation's monthly calendar showed even more intense localism, as 89 percent of advertisers fell within the parish boundaries.

By 1985, the publication of the St. Thomas centennial volume, the distances between businesses and the church had increased modestly. On average advertisers remained less than a mile from the church. However, a far smaller proportion of advertisers, 26 percent, were located within the parish. Although 68 percent remained within the bounds of Italian South Philadelphia, localism had passed its peak (map 8). Notably, 13 percent of ads represented businesses in the suburbs, a group that had never claimed more than 1 percent of ads in previous volumes. The movement of St. Thomas parishioners to the suburbs meant that businesses there could envision a connection to South Philadelphia. These more distant business owners may have purchased ads because of their own connections to the church—indeed they may have resided in the parish but operated businesses in the suburbs—but they likely also expected that suburban-dwelling Italian ethnics would return to South Philadelphia for the one hundredth anniversary of the parish and purchase the memorial volumes. A quarter-century earlier, for the parish's seventy-fifth anniversary, only 1 of 141 advertisers had followed a similar logic. Still, distant advertisers remained a small minority of the total, even in 1985. Most advertisements reflected the connections among church, commerce, and community within the geographic boundaries of Italian South Philadelphia.

In Toronto, meanwhile, the spatial contours of parish advertisements had a different character. In both 1959 and 1978, two of three

MAP 7 Locations of advertisers in St. Thomas "Diamond Jubilee" parish publication, 1960. SOURCE: "Saint Thomas Aquinas Parish Diamond Jubilee," 1960, box 48, group 102, Parish Histories, PAHRC (138 addresses displayed).

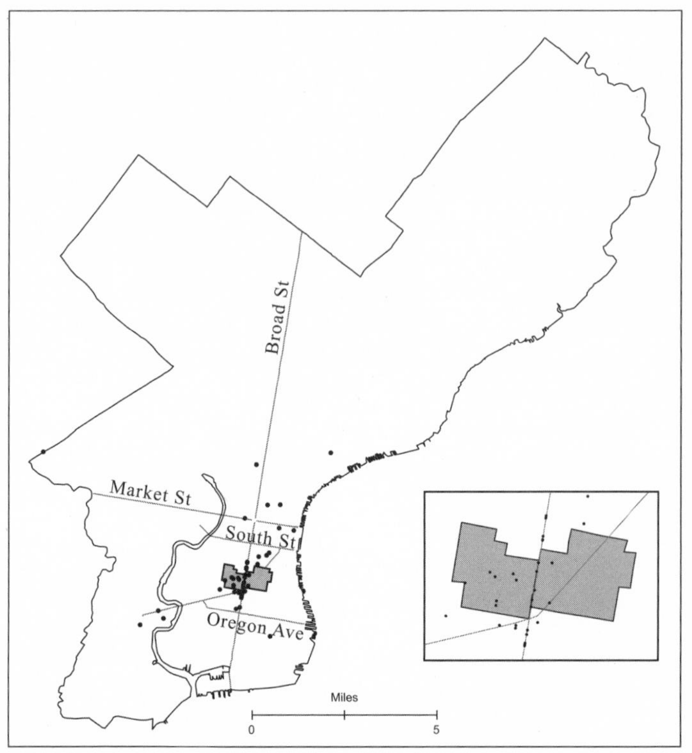

MAP 8 Locations of advertisers in St. Thomas Centennial Volume, 1985. SOURCE: "St. Thomas Aquinas Parish, 1885–1985," box 48, group 102, Parish Histories, PAHRC (51 advertisments displayed).

advertised businesses were located outside of the parish.[89] In Toronto, the Italian social networks that brought advertisers to the volume did not require locality; the importance of propinquity declined over time while that of ethnicity increased. In the first volume, published in 1959, advertisers were scattered across metropolitan Toronto. Although the area around the parish contained more advertisers than any other, the volume included businesses located throughout the northwestern part of the city (map 9). Their average distance from the church was 2.3 miles, six times further than the geographic spread of the Philadelphia advertisers at similar dates (fig. 26). Twenty years later, the difference between South Philadelphia and Italian Toronto had increased. On average, 5.7 miles separated St. Francis Church from the advertisers in the parish's seventy-fifth anniversary volume, and more than two-thirds of advertisers were located more than four miles away from the church (map 10). The trend emerging in South Philadelphia by the late 1980s was characteristic of Little Italy from the start—advertisers were connected to the parish over long distances.

Italian business owners in Toronto thus expressed ongoing confidence in the elasticity of ethnic social bonds. Some may have expected parishioners living in Little Italy to travel to their businesses, while other advertisers—for example, a laundromat, a photo studio, and a confectioner in 1959; a donut shop, a bakery, and a tavern in 1978—likely could not expect to draw from a wide radius. Instead, their ads reflected a belief that the parish, and its memorial volumes, would appeal to people living outside of Little Italy close to their own businesses. In 1959, such bonds usually ended with metropolitan boundaries, but twenty years later the volume offered evidence that this boundary too was giving way. The proportion of advertisers outside of Metro Toronto altogether increased from less than 1 percent in 1959 to 17 percent in 1978. The new Italian areas at the northwestern edge of the city were increasingly connected with the parish in Little Italy. Elastic Italian ethnic bonds extended across metropolitan Toronto and sometimes beyond.

Crucially, the volumes not only evidenced but also reinforced the

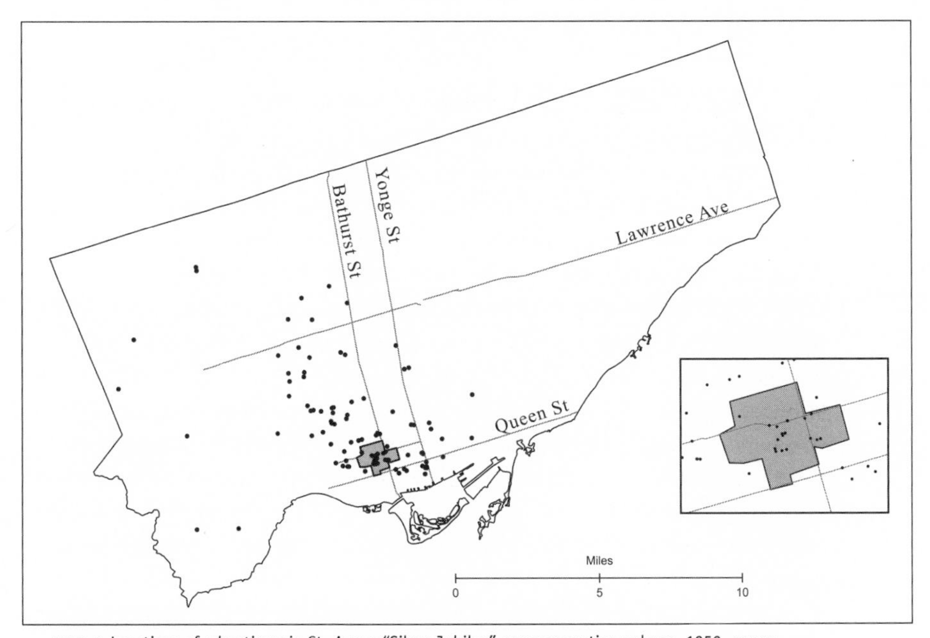

MAP 9 Locations of advertisers in St. Agnes "Silver Jubilee" commemoration volume, 1959. SOURCE: "Silver Jubilee Commemoration at Saint Agnes Church, Grace and Dundas Streets Toronto, Canada," Parish History: Franciscan's Twenty-fifth Anniversary Publication, 1959, St. Agnes Parish Papers, ARCAT (103 addresses displayed).

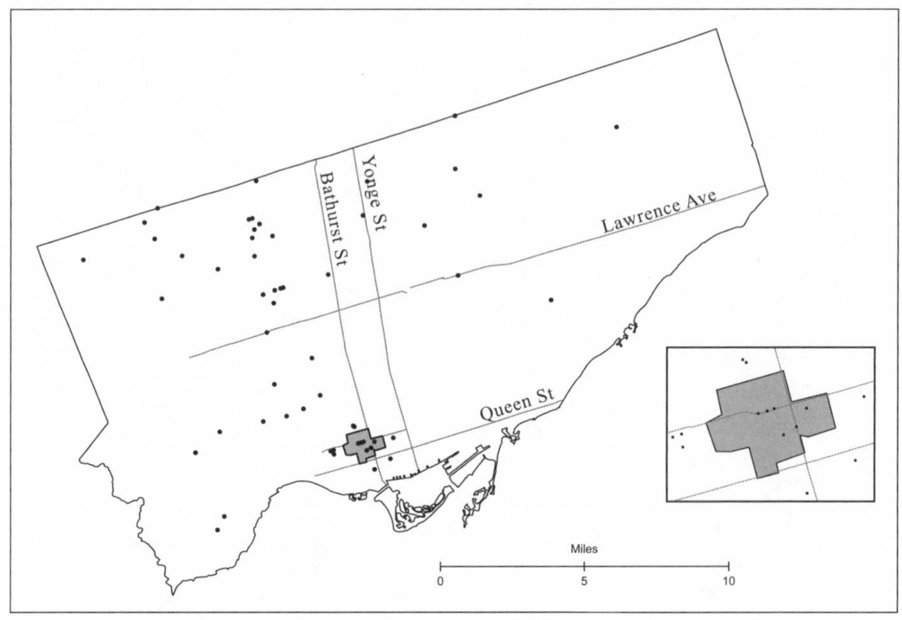

MAP 10 Locations of advertisers in St. Francis "Diamond Jubilee" volume, 1978. Source: "Diamond Jubilee Celebration, St. Francis/Old St. Agnes Parish Community, 1903–1978," Publications: Seventy-fifth Anniversary Book, St. Francis Parish Papers, ARCAT (61 advertisements displayed).

geographic dynamics of their respective parishes. Parish volumes in Italian South Philadelphia reinforced localism, while those in Little Italy encouraged elasticity. Parishioners and others who purchased the volumes in Philadelphia found themselves in possession of very different documents than those produced in Toronto. At Annunciation and St. Thomas, memorial volumes reminded readers of the goods and services available to them within Italian South Philadelphia, whereas Italian Torontonians left church with a guide to Italian businesses throughout the metropolitan area. Thus South Philadelphia's memorial volumes reinforced parish boundaries—separating one part of the city from the rest—while those in Little Italy functioned as an institutional mechanism connecting Italians dispersed across the city with one another. Ethnicity, commerce, and church participation ran together in both places, but the bonds among these facets of life left very different urban footprints.

As with exchanges of housing, religious practices in Toronto's Little Italy and Italian South Philadelphia diverged in spatial form. With the important, and ironic, exception of the *Festa Italiana* in St. Thomas Parish, all evidence suggests that parish life in Italian South Philadelphia and Toronto's Little Italy was highly ethnically endogamous. Italians supported the church through advertisements in parish publications, and by and large, they were the exclusive participants in the wide range of parish activities, including processions. However, the spatial dynamics of Italian ethnic connections and their institutional manifestations differed in the two contexts. In Italian South Philadelphia, localism shaped institutional life. Vibrant church activities were often exclusive to parishioners; outsiders who did attend church events in St. Thomas and Annunciation usually came from one of the other Italian churches in close proximity. Parishes were thought of as neighborhood institutions, even if neighborhood was sometimes imagined to include non-Italian local residents. In Toronto, by contrast, a territorial logic did not prevail. The church of Little Italy served Italians in the entire metropolitan area and beyond. The parish in Little Italy fostered ethnic bonds by inviting Italians to a shared space at the city center. Church events mirrored the exchange of real estate because both reflected the efforts of Italian ethnics to negotiate their particular urban environments. At the same time, economic and associational practices reinforced one another. Networks forged at parish celebrations and in church halls served economic functions and facilitated further economic exchanges. As these varied activities converged to shape social experience, very different kinds of ethnic community emerged.

4 Courtship, Marriage, and the Geography of Intimacy

Anthony S. was in his mid-twenties when his future wife, Elizabeth, moved to his block on Camac Street in South Philadelphia. Although she had previously lived only a short distance away, the two met for the first time after becoming neighbors: "I knew her brothers . . . and I kept saying, 'Hey, I like your sister, I like your sister.'" Eventually Anthony approached Elizabeth with a similar message and the two were soon an item. A typical date consisted of a drive through the neighborhood:

> I used to have a little blue Chevy with two little dolls in the back seat . . . and I used to put holes in my muffler with an ice pick so you could hear "Blub-blub-blub-blub" . . . And we'd just drive around, wave to everyone, "Hey I got a car, how ya' doin'?" And we used to wave to everyone up and down Thirteenth Street, Twelfth Street, whatever street we'd go. Broad Street was the main thing; you'd just wave to everybody.

The relationship, so contained within the neighborhood, was threatened when Anthony departed South Philadelphia for military service in Korea: "And then I got called in the army, and she says, 'We going out?' And I says, 'No, we're not going out no more because I'm in the army . . . I ain't going to worry about you 'cause I'm going to Korea.'"

FIGURE 27 Anthony S. and Elizabeth, newlyweds, depart for their honeymoon in 1957. By permission of Anthony S., personal collection.

But neighborhood ties were not so easily severed. Even with Anthony gone, "She'd always go to my mother's house when I was away and help her, she lived right up the street . . . She used to come to the house to help bake cakes and stuff with my mother." When he returned from the service, the couple resumed dating. Eventually, Anthony thought, "Gee, it's over a year we've been going out together, I'm getting old, what are you going to do?"[1] Overcoming the objections of Elizabeth's mother, who preferred a local undertaker's son, the couple wed in 1957 (fig. 27).

Young people in Toronto and Philadelphia made choices in marriage that reflected and reinforced the broader social patterns of their neighborhoods. Marriage, though not an everyday choice, grows from a much wider array of activities, social surroundings, and local practices. Anthony and his wife met because they lived on the same street. The pair met as neighbors, brought together by siblings and parents, and their romance was nurtured on the streets of the parish. In South Philadelphia, as another parishioner put it, "You all knew their mothers and fathers and aunts and uncles, you didn't have to be introduced."[2] Marriage choices in Italian South Philadelphia and Toronto's Little Italy—and the differences between those choices in each context—reveal the personal impact of urban economic and institutional structures. Young people seldom see urban change as a factor in their romantic choices. As Anthony S. laughingly recalls the girl in his gurgling blue Chevy, he mentions nothing in the way of racial division or housing markets. Nonetheless, his choice in marriage—like those of other Italians in Toronto and Philadelphia—demonstrates the importance of urban environments to even the most intimate decisions.

At the same time, small, personal choices sustained larger urban patterns. Italian South Philadelphia and Toronto's Little Italy maintained their particular social characteristics—ethnic homogeneity and residential stability in Philadelphia and ethnic heterogeneity and geographic mobility in Toronto—because of the choices of young people. Housing exchanges and institutional arrangements were inextricably interwoven with corresponding practices of marriage and household formation.

In both Toronto and Philadelphia, ethnic community was sustained by Italians who chose to marry other Italians. Both the descendents of Italian immigrants to South Philadelphia, Anthony and Elizabeth were representatives of a consistent pattern in postwar urban ethnic neighborhoods. Even as ethnic intermarriage rates in both Canada and the United States soared, in pockets such as South Philadelphia and Toronto's Little Italy, Italian social networks continued to wield powerful influence on marriage choices.[3] One Italian Torontonian put the formula simply: "You belonged to your dances or your clubs you went to, you married the ones that you met."[4] A parishioner in St. Thomas, describing social circles in Italian South Philadelphia, reflected, "They were all Italians . . . I don't think they mixed with anybody."[5]

Catholic marriage registers confirm the importance of ethnicity to romantic choices in Italian South Philadelphia and Toronto's Little Italy. Three kinds of information in the registers—names, places of baptism, and residence at the time of marriage—provide a window into the rooms where young people met. If housing and institutional histories tell us much about these rooms—how they were exchanged and organized—marriage registries tell us who gathered inside of them. The romantic connections among young people in St. Agnes/St. Francis in Toronto and Annunciation and St. Thomas in Philadelphia illuminate, in greater detail than any other source, both the similarities and differences between the social networks of Toronto's Little Italy and Italian South Philadelphia.[6]

Italian brides in both contexts showed appreciation for men of Italian origins. In 1950, almost nine in ten brides in Annunciation and St. Thomas Parishes bore an Italian name.[7] Among these, nine of ten married an Italian man (figs. 28 and 29). The numbers remained steady through the 1970s before dropping in 1980 and 1990. Yet even in 1990, when diverse new immigrants to South Philadelphia made the cohort of brides less homogeneous and the steady depletion of the Italian American population encouraged those remaining to seek other partners, the majority still took a husband of Italian extraction. The local Italian networks that brought brides and grooms together in South Philadelphia were on the decline, but far from extinct.

In Toronto, the pattern of marriage changed differently over time, but as in South Philadelphia, women in Little Italy appreciated men with Italian names. In 1950 almost eight in ten women with Italian surnames married men with the same background, a rate just below that of their South Philadelphia peers (fig. 29 above). With the onset of immigration, the number shot upward; in the 1960s and 1970s all but a tiny proportion of Little Italy's brides took an Italian partner.

During this period, localities and hometowns of origin steered the romantic choices of Italians in Toronto. In 1960, two-thirds of immigrant brides married someone from their own Italian region of origin,

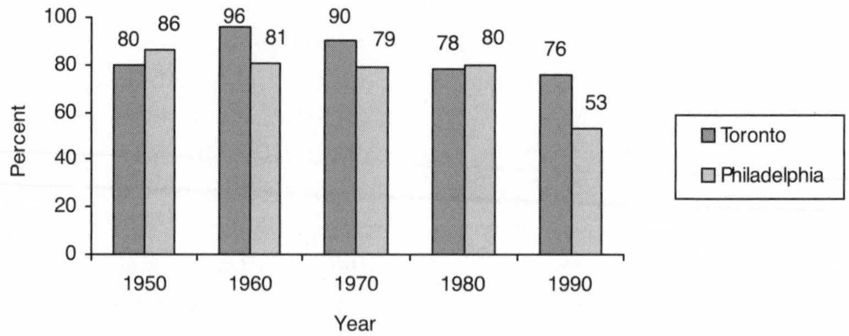

FIGURE 28 Percentage of brides with Italian surnames, Toronto's Little Italy and Italian South Philadelphia, 1950–1990.

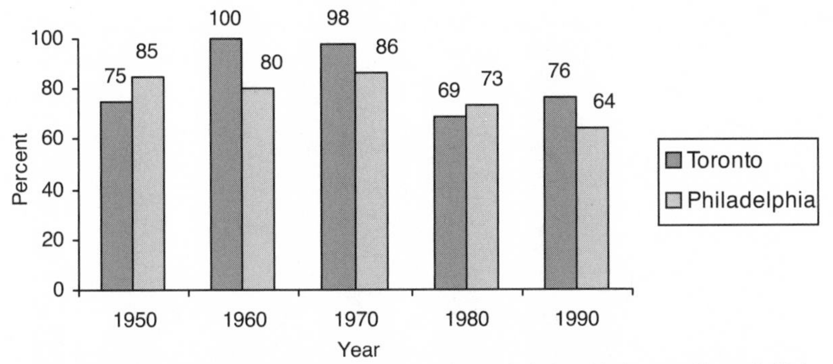

FIGURE 29 Percentage of brides with Italian surnames who married men with the same, Toronto's Little Italy and Italian South Philadelphia, 1950–1990. SOURCE: Marriage Register, 1950, 1951 1960, St. Agnes Church, ARCAT; Marriage Register 1970, 1980, 1990, St. Francis Church, ARCAT; Marriage Register, 1950, 1960, 1970, 1980, 1990, Annunciation and St Thomas Churches, Philadelphia. $N = 1021$ (Philadelphia 721, Toronto 300).

and more than half took a groom from an even smaller home-country geographic area, the provinces, which encompassed a city or a number of towns (fig. 30). The figures for Calabria, the most common regional origin of immigrants to Little Italy, were even higher, suggesting that the region and Cosenza, its primary province of embarkation, constituted their own social circles in Toronto.[8] Francesca S. recalls the bonds and antagonisms that encouraged marriages among people from the same localities. Sicilians in the eastern section of the city where she lived retained business and social ties with Sicilians in the College Street area, and "many marriages happened . . . between the people living in College and the Sicilian in the East End . . . the family knew each other from before, from many years ago . . . so there was a link."[9] Meanwhile, marriages that spanned different home regions were often discouraged: "A [Sicilian] father wasn't happy when the daughter told him that she was in love with a Foligno guy . . . might as well say you are marrying an Austrian or a German."[10] Having faded in the prewar period, these strong local attachments, known among Italians as *campanilismo*, reemerged in the context of immigration.[11]

However, regional ties and tensions declined over time as they had after the previous wave of Italian immigration to the city. In 1970, the vast majority of Italian-born brides continued to marry Italian-born grooms, doing so in 95 percent of cases. However, just over half chose a groom from the same region, and less than half married a groom from the same province. By 1980, with the Italian-born pool declining, brides born in Italy often married Canadian-born men with Italian surnames. Vince P., who settled in Canada with his parents in the early 1960s,

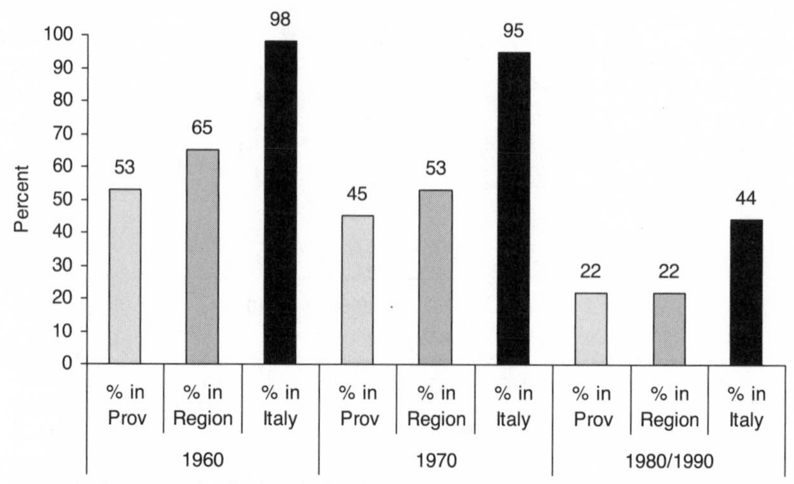

FIGURE 30 Percentage of Italian-born brides whose grooms come from the same place, by province, region, and country, Toronto's Little Italy, 1960–1990. SOURCE: Marriage Register, 1960, St. Agnes Church; Marriage Register, 1970, 1980, 1990, St. Francis Church, ARCAT. 1960, *N* = 40; 1970 *N* = 38; 1980–1990 *N* = 18.

remembers intergenerational conflict and accommodation over *campanilismo*. As he recalls, "parents realized that things were not gonna happen the way they thought they were gonna happen." Immigrant parents came to accept marriages that crossed Italian regional boundaries. Whereas the marriage of a child of Calabrese immigrants to a partner from the Abruzzi or Sicily initially caused "shock for the parents," by the mid-1970s, such liaisons had become increasingly acceptable.[12]

Thus, as it had a half-century earlier, postwar Italian ethnicity in Toronto initially bore the mark of the regional networks that facilitated immigration itself, but as immigrants from various regions intermingled in Toronto, regional networks faded from dominance.[13] While some associations—such as the Loggia Pisticci, founded at St. Agnes—encouraged ongoing connection to locality of origin, other societies, the church, schools, and neighborhood life brought Italians of diverse origins together, and such gatherings bore fruit.

Although women in both parishes found Italian spouses, over time the geography of these connections diverged in the two locales. Whereas Italian South Philadelphians found local spouses throughout the postwar period, the brides of Toronto's Little Italy looked increasingly far from home. The geography of intimacy followed, and reinforced, the patterns laid down in the rest of social life.

However, this difference between the two neighborhoods emerged gradually. In the early postwar period, brides in both Italian South

Philadelphia and Toronto's Little Italy typically found spouses in their local neighborhoods. Although those in Toronto's Little Italy already looked slightly further afield, in 1950 this difference, only 0.7 miles on average, likely lacked real social significance (fig. 31). Similarly, in 1960 brides in both areas found spouses living relatively close, on average of less than 2 miles, from their homes. Thus, at the outset of the postwar era, young people in both places connected with spouses through the local bonds that had characterized prewar ethnicity (maps 11 and 12).[14] In the 1950s and 1960s young men and women in both locales met in rooms crowded by neighbors.

Residents of both areas recall the importance of neighborhood during the early postwar period. In South Philadelphia social ties were fostered by robust street life. Especially in the summers—when the oppressive heat forced residents out of their row houses and onto their front steps—South Philadelphians came to know one another on the narrow streets of their neighborhood. Cathy S. recalls summer evenings on Passyunk Avenue, the main commercial strip in Annunciation Parish, where neighbors gathered after the shops had closed (fig. 32). Cathy remembers, "everybody would sit out. All the neighbors would be out there. The old women would be out on their beach chairs and the kids would be out on the step." She and her father both gathered with friends in front of the store that he operated on the first floor of

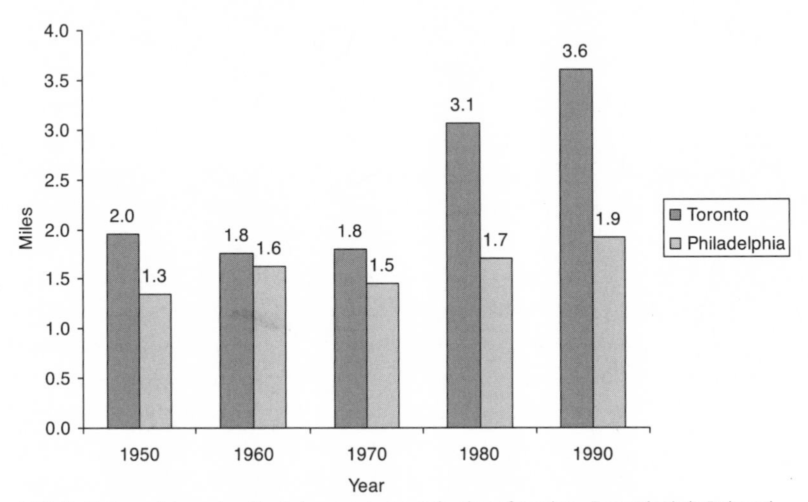

FIGURE 31 Mean distance in miles between spouses at the time of marriage, Toronto's Little Italy and Italian South Philadelphia. SOURCE: Marriage Register, 1950, 1951 1960, St. Agnes, ARCAT; Marriage Register, 1970, 1980, 1990, St Francis, ARCAT; Marriage Register, 1950, 1960, 1970, 1980, 1990, Annunciation and St. Thomas Churches, Philadelphia. *N:* 775 (Philadelphia 558, Toronto 217).

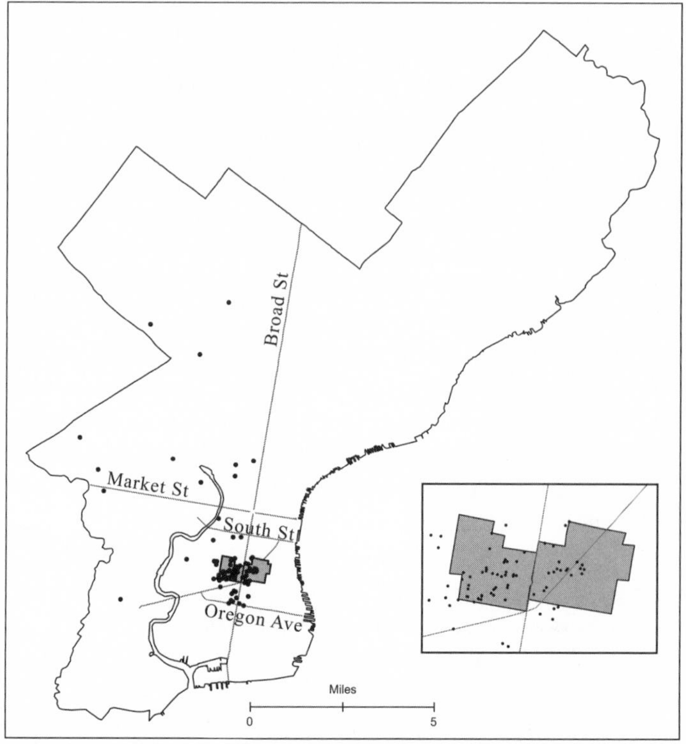

MAP 11 Addresses of grooms married in Annunciation and St. Thomas Parishes, 1950. Marriage Register, 1950, Annunciation and St Thomas Churches, Philadelphia (100 grooms displayed—50 per parish).

their house: "[W]e had a bench out there. And he used to sit out there with his friends . . . we would be out there playing cards."[15] Anthony S. describes the social landscape on his corner, several blocks away from where Cathy sat with her friends:

> We used to have . . . not a gang, a clique . . . six, seven, eight guys used to hang on the corner. And on our corner, on Twelfth and Morris, we used to be three different kinds of gangs or cliques. One was the old, the older one, they used to call them the forty thieves . . . they were thieves, they used to steal things different places, and they would hang on the corner after. Then there was the brothers, like my brother . . . My crowd, my age, all had older brothers . . . we'd get the rotten tomatoes and stuff and try to hit them with it, then we would have to run away.

According to Anthony, the "cliques" were a welcome part of neighborhood life. "Old ladies" in the neighborhood, he recounts, "couldn't wait for us to hang on the corner . . . they knew they were safe . . . if they come home with a package one of us would get it and walk it to the house . . . We always took care of the neighborhood."[16]

Annunciation parishioners recall that neighborhood ties encouraged romantic connections. Cathy S. remembers local boys intruding on her card games: "The guys would come over and bother us: 'Don't play that card. What are you doin'?'" Another parishioner describes women meeting young men "hanging on the corner" outside local stores. Her longtime friend described herself as a typical example: "I lived on this corner, and if you walked to the next corner was my husband, and we got to know each other and we went out." One Annunciation bride met her husband at her family business: "His mother used to come in and buy the groceries, and he would come in once and awhile . . . we started going to skating parties and the dances and all, and that was that."[17] When Anthony S. and his friends grew too old for tomato fights, they left their street corner to linger in the back booths of the "blue room," a local luncheonette, hoping to meet local girls: "Anytime you get the girls

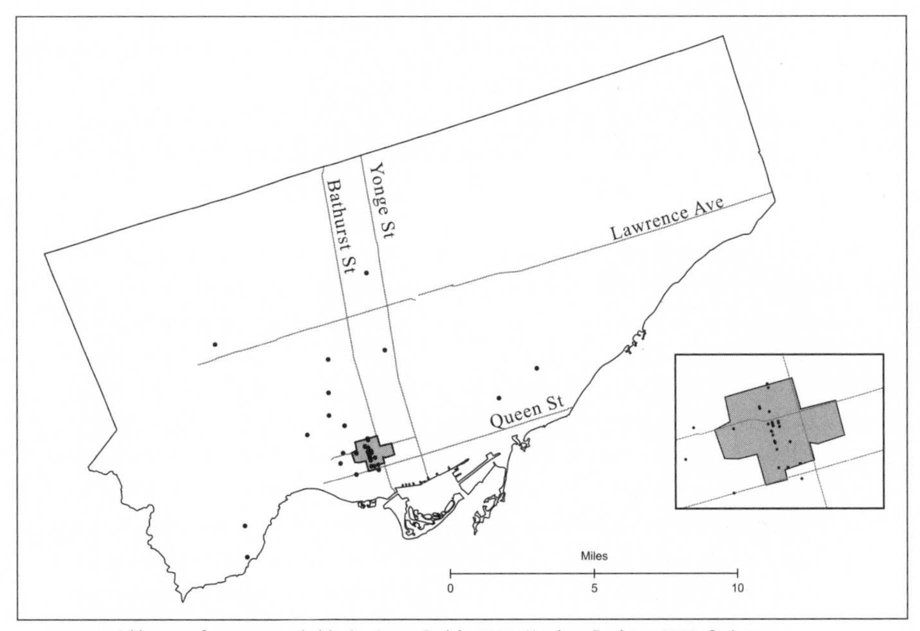

MAP 12 Addresses of grooms married in St. Agnes Parish, 1950. Marriage Register, 1950, St Agnes, ARCAT (40 grooms displayed).

FIGURE 32 Shopping on Passyunk Avenue in Annunciation Parish, 1980. Source: *Philadelphia Evening* Bulletin, Photojournalism Collection, Streets-Passyunk, May 18, 1980. Philadelphia, Temple University Archives, Urban Archives, Philadelphia, PA.

come by—'Hello, hello, hello'—you would sit out in front with them, you buy them a soda."[18]

Recollections of Toronto's Little Italy in the early postwar era also suggest the importance of neighborhood ties. An Italian immigrant who lived near the intersection of Manning and College Streets remembers that "everybody knew everybody, everybody visited each other and on a Saturday night it was like a street dance."[19] With immigrants flooding the marriage rolls in 1960 and 1970, the neighborhood was a place where newcomers could expect to be understood in stores, streets, and bars. One immigrant to the city in this era recalled: "During the evening when I was free I was always on College Street, where a lot of Pisticesse used to get together, by hundreds. College Street was dominated by Pisticesse."[20] Another migrant, from Siderno, Calabria, also thought of the street as his own: "College Street was my . . . my street. We had a bunch of friends there, we used to roam around."[21] Another remembered: "We used to walk up and down, look at the girls and vice versa . . . just killing time because there was no money to do anything else."[22] Vince P., also a resident of the area in the 1960s, recalls that the gentle curve in College Street near Grace, in the heart of Little Italy, and the rumbling streetcars slowed local traffic, permitting the flourishing of a pedestrian culture that mimicked the piazzas in Italian towns: "The pleasure was in the walking in the street and street meetings. You went for coffee and you went for a long walk . . . College Street became a meeting place."[23] The area gave recently arrived immigrants a sense of place in the city. Gathering on the streets, local residents reinforced Italian ethnic ties as well as the connections between emigrants from the same Italian localities (fig. 33).

In these regards, Italian South Philadelphia changed little in the decades that followed. As the area's population declined, Italian South Philadelphians closed ranks. Patterns laid down in the 1950s continued into the 1970s and 1980s. Italian South Philadelphia remained an intensely local place. In 1950, most women married at Annunciation and St. Thomas were lifetime residents of the area. Four out of five brides had been baptized at a church in South Philadelphia. The churches were located close to one another, so that by the time of their marriages, despite often moving to a different parish, the young women still lived only a short walk, on average less than a mile, from their places of baptism.[24]

FIGURE 33 A crowd gathers to watch a film crew outside iconic Italian businesses on College Street, 1975. Photograph courtesy of Vincenzo Pietropaolo.

A generation later, brides in the parish resembled their predecessors. Among those who married in 1980, two-thirds took their First Communion at a church in South Philadelphia, and the mean distance between their residence prior to marriage and their places of baptism remained less than one mile.[25]

Remaining within a small geographic radius encouraged close relationships among lifelong South Philadelphians. Cynthia C. moved only once in her seventy years of life, relocating with her parents at the age of thirty from the house in which she was born to a larger one across the street. Cynthia recalls, "Your neighborhood was an extension of your family . . . you seemed to stay in the same neighborhood for years . . . you not only had your mother and father, but you had neighbors that looked out for you."[26] Anthony S. also describes neighbors as care givers, "somebody gets sick you always got these people around . . . they'll take care of everything."[27] In St. Thomas Parish, Ida G.—resident of the same house for her entire eighty-seven years—boasts of her enduring connection with her lifelong neighbor, John, whose birth in the adjoining row house was assisted by Ida's mother eighty years prior.[28]

Brides in Italian South Philadelphia expressed their rootedness in place by continuing to choose local grooms. In 1980 and 1990, brides married men who lived, on average, within two miles of their homes. However, marriage networks increasingly crossed parish boundaries. In 1950, almost one-third of marriages involved brides and grooms residing in the same parish. Twenty years later, less than 15 percent did the same. The pool of marriageable candidates within any of the shrinking individual parishes grew insufficient; to find a partner, young people had to look further from home. So, prospective brides and grooms moved in a somewhat larger area. They met at Bishop Neumann High School or at the shops and ice cream parlors on Broad Street and Passyunk Avenue that served all of Italian South Philadelphia. By 1990, some 90 percent of Italian South Philadelphians married outside their own parish, but the majority of brides still married a South Philadelphian.[29] Despite breaking free of their parishes, many South Philadelphians continued to restrict themselves to the limits of the neighborhood (map 13).

The persistence of localism in South Philadelphia defied demographic odds. The population of the census tracts in Annunciation Parish and the eastern half of St. Thomas declined by more than 50 percent from 1950 to 1980. Further, older residents were most likely to remain behind. The proportion of people in their twenties and early thirties within the local population fell by a third between 1950 and 1980, while the proportion of residents over sixty years of age more than doubled.[30]

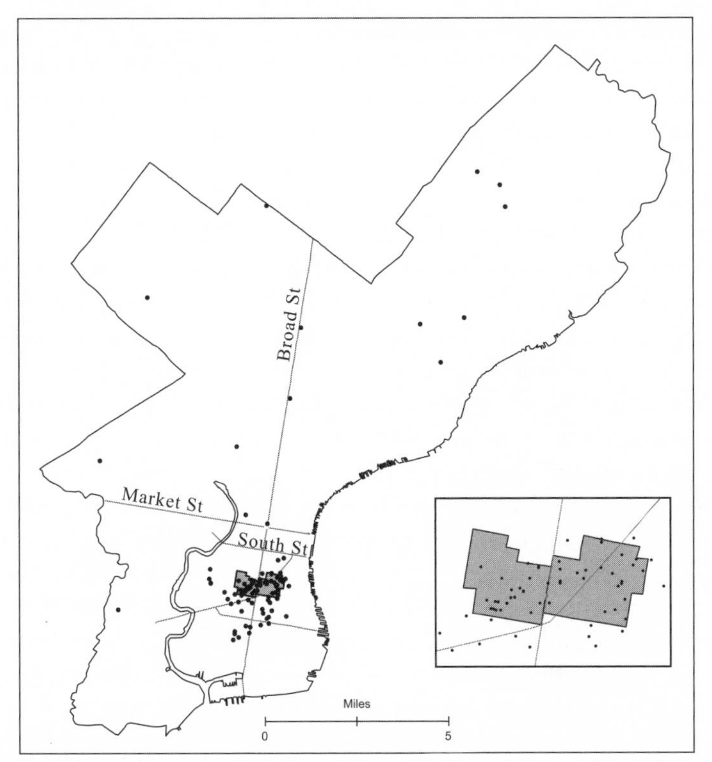

MAP 13 Addresses of grooms married in Annunciation and St. Thomas Parishes, 1980 and 1990. Marriage Register, 1980 and 1990, Annunciation and St. Thomas Churches, Philadelphia (100 grooms displayed—50 per parish).

Young people were departing from South Philadelphia en masse, leaving fewer local choices for women who remained behind. Intimate social networks might have been expected to expand to reflect this shift. As young Italian Americans spread across the suburban frontier, the lines connecting brides to their grooms might have been expected to follow. Instead, the geography of social life remained fixed; brides continued to find spouses close to home. This rule held true at the periphery of Italian South Philadelphia, in St. Thomas, as well as in its core, in Annunciation.[31] By scarcely increasing the geographic scope of their social networks, young people fit the pattern set down by homeowners seeking to grow old in the company of their friends and conformed to the shape of local institutional life.

In Toronto, local ties proved less enduring. By the 1970s, College Street was but one of several centers of Italian life in Toronto and the wider region. Italian immigration to the city poured across old boundaries into the large northwestern expanses of the metropolitan area. Fruilani immigrants, from northern Italy, established a club on Brandon Avenue north of Little Italy. Further north, St. Clair Avenue emerged as a center of vibrant Italian life. Meanwhile, Italians moved out of the College Street enclave. As previously noted, the Italian origins of the population in the tracts encompassing the Italian parish fell from almost 12,000 in 1971 to just under 5,000 ten years later. As in the Philadelphia region, then, people of Italian origins were spreading out. Unlike in Philadelphia, however, the residents of the oldest Italian enclave in Toronto developed intimate, enduring connections with Italians throughout the metropolitan area.

The Italians in Toronto's Little Italy had less personal investment in their neighborhood than their peers in South Philadelphia. They were, as a group, far less rooted in place. In 1960, nine in ten brides at St. Agnes had taken their first rites in Italy; ten years later, the number declined very little to eight in ten. While Italian South Philadelphians lived within social circles generations in the making, the new immigrants in Little Italy were accustomed to friendships that could emerge serendipitously. Paul N. made his first Italian Canadian friend en route to Halifax: "On the boat was very, very good for me . . . I didn't get sick and I made a friend . . . and we remained friends all our lives . . . we were having fun while our parents were in their cabins sick."[32] If some bonds emerged during travel, others were stretched long distances due to relocation. In 1959 Vince P.'s entire family moved from Calabria to Toronto, but almost two years later—with his father struggling to find steady employment—all except his father and eldest brother returned home to their village. The family was reunited in Toronto some two years later.[33] In the process of migration, families, friends, and lovers learned to maintain ties across vast distances.[34] Immigrants to Toronto understood that local persistence was not a prerequisite of intimacy. The Canadian-born children of immigrants also understood this history of motion.

In the 1970s, the neighborhood ties binding social life began to give way in Toronto as young people in Little Italy sought spouses further from home. By 1980, the couples married in Toronto's Little Italy met across a significantly greater distance than their peers in South Philadelphia. In Philadelphia's Annunciation and St. Thomas Parishes, 64 percent of brides and grooms continued to reside within a mile of one another. In St. Francis Parish in Toronto, only 36 percent did the same. While the average distance between spouses in Toronto's Little Italy steadily rose, brides in Italian South Philadelphian continued to marry

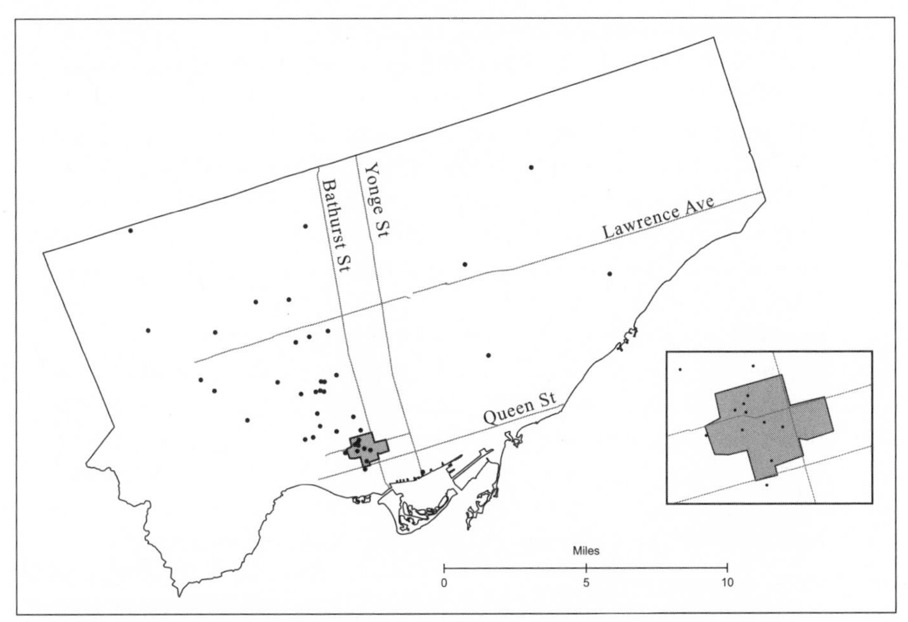

MAP 14 Addresses of grooms Married in St. Francis Parish, 1990. Marriage Register, 1990, St Francis, ARCAT (42 grooms displayed).

men living, on average, less than two miles from their homes. In the years that followed, this difference between the Italian areas increased. By 1990 the couples married in Toronto's Little Italy were separated by almost twice the distance of their South Philadelphia peers (map 14).

The flow of Italians through Toronto, unlike the well-articulated experience of Italian neighborhood, has received little scholarly and popular attention. Metropolitan networks are little emphasized in most descriptions of social life. But evidence that Italian immigrants could move widely across Toronto slips into accounts of postwar experience. Paul N. recalled that as teenagers in the 1960s, he and his friends would travel east of Little Italy to Yonge Street, where they could spend a weekend evening simply driving up and down the city's central business strip: "I remember we used to drive up and down . . . for what reason I don't know . . . because that's the thing to do."[35] Another Italian immigrant, Antoniette Ranieri, described her social wanderings in detail:

> [W]e used to walk from Dufferin to Yonge Street . . . from Dufferin to Eatons, downtown . . . we were used to walking . . . we used to [go to] Yonge Street and watch the parades like the

> Knights of Columbus used to have their parade, and if the boys were in it we'd go with them . . . to watch the boys play ball.[36]

Movement across the city became an increasingly routine aspect of social life for postwar Italians in Toronto. They traversed space casually. Social life was not territorialized in the fashion of postwar Philadelphia.

Some immigrants specifically recall travelling to Little Italy during their leisure time. Vince P. lived in Little Italy during the 1960s but moved northwest to St. Clair in 1974, and then further north again to Eglinton and Dufferin. While these areas also included notable Italian contingents, he often returned to Little Italy: "I would actually come down to the area just to have a coffee here. It was important to sit down to have a cappuccino, have an espresso on College Street." While Italian coffees were available in his new neighborhoods, "it wasn't the same ambiance as it is here . . . Here you have a coffee and you look at it for a long time, you savor the feeling of the coffee, the coffee is a pretext . . . let's go for coffee means let's get together."[37] Francesca S., who never lived in the area, fondly recalls family outings from the Danforth, in the eastern end of Toronto, to College Street Little Italy in the 1970s. She and her husband would gather their children into their car and set out to Little Italy in search of Italian products: shoes, chestnuts, ice cream—often items that they could find closer to home, but instead chose to purchase in the city's oldest Italian enclave. "Going to College was an outing," she explains; it was "the traditional escape." At the Sidewalk Café, a popular ice cream parlor on College, Francesca's husband could speak Sicilian with the owner while their children licked their gelatos.[38]

Despite the emergence of other centers of Italian life in the region, Little Italy remained a key gathering place. Those who had left the area came back. Italians who had never resided there visited nonetheless. These journeys brought young people together. By the 1970s and 1980s, the rooms where the youth of Little Italy sought romance had shifted significantly. Whereas previous generations had gathered with neighbors, the new geography of intimacy tied together the Italians of the metropolitan area.

Some accounts of travel suggest that the availability of good, cheap public transportation facilitated the movement of Italians across the city. Frank S. recalled that on the day of his arrival in Toronto in the late 1960s he found himself lost, standing on the corner of Dundas and Yonge Streets, map in hand: "Somebody stopped me and said, 'Where do you want to go?' . . . told me to take the subway."[39] Vittorio Zavagno and his wife took the streetcar every Saturday from Little Italy to the St. Lawrence market, where they would purchase their groceries for the week.[40] As a child, Vince P. spent a summer wandering around the western half of the city. Although he did not ride the streetcar, their tracks served as a trail guide: "I didn't speak English and I didn't know where I was going, and I'd follow streetcar tracks and figured as long as I follow the streetcar tracks, I'll never get lost." In the years that followed, Vince's wandering continued. A burgeoning interest in photography led him across the city: "I would be exploring the city . . . I'd be going on long walks far away with my camera, my camera would be a tool for exploration." Such excursions were possible, on Vince's account, because the city was understood as a "very safe place" where nobody feared for a boy wandering for hours by himself.

Decades later, Mirella B. told a particularly remarkable story of her streetcar travel in the mid-1960s. After learning that her husband had injured his hand and been admitted to a hospital on the eastern edge of Toronto, Mirella, then seven months pregnant, set out across the city in the darkness of night:

> Since I couldn't speak a word of English I give him [the streetcar driver] the little note with the address . . . I had to change . . . he actually stopped: he took me by the hand, he took me down the three steps of the streetcar . . . he waited there until a bus arrived and then he talked to the driver of the other bus.[41]

A young Italian woman, recently arrived in Canada and very pregnant, would seem an unlikely candidate for a long trip across a strange city at night. But the patterns of social life in Toronto's postwar Italian community made such a trip, if not commonplace, at least conceivable. Good, cheap public transportation helped Italians move through Toronto. The streetcar that carried Mirella likely brought many grooms to their Little Italy brides. But use of public transportation in this fashion also required that the social organization of city space facilitate fluidity. Conduits, rather than boundaries, dominated the social landscape of metropolitan Toronto.

The freedom of young women from Little Italy to connect with men from across Toronto reflected not only the spatial dynamics of Italian ethnicity but also the changing contours of gender. According to one older Italian Torontonian, before World War II she and her peers in St. John's Ward were strongly expected to marry local men: "If you married outside the district it was a crime . . . your mother and father had to know the family . . . There was no outside."[42] During the postwar era, social ties ceased to be limited by locality, and the spatial freedom of young Italian women greatly increased. In South Philadelphia, by contrast, the social lives of young women continued to be constrained by locality. The social geography of marriage in Toronto included women empowered to move about the city in their leisure and social lives, while in South Philadelphia women were spatially constrained.[43]

Marriage choices in Italian South Philadelphia and Toronto's Little Italy followed the distinct patterns of ethnicity in each city. Although Italians came together in both cities, they did so differently. In Philadelphia's Annunciation and St. Thomas Parishes, marriage ties depended on networks rooted in locality; intimate connections formed as a result of propinquity. In Toronto, by contrast, marriage networks showed the elasticity visible in the rest of social life; ethnicity stretched far and wide across the metropolitan area.

Marriages tie the practices observed in the housing markets and associational life together with the experiences of individual parishioners. They illustrate the ways in which different spatial practices in broader social life could reverberate in the decisions of young people. In South Philadelphia, housing practices encouraged a stable neighborhood that linked spouses to one another on the basis of locality. Residential flux in prospering Toronto brought new faces into the area around the church while dispersing Italians throughout the city—ethnic networks adapted. Institutional practices are similarly implicated in the choices of young people. Differences between the geography of church processions and social associations helped to shape the choices that young people made in marriage.

But marriage choices also sustained the wider patterns of social life. Young people forming new households and families populated real estate markets and church pews. Their decisions breathed life into social networks; their activities constantly reconstituted social life over time. The patterned differences between their choices set ethnic life in Toronto's Little Italy and Italian South Philadelphia on distinct postwar trajectories.

Of course, not every Italian South Philadelphian found her spouse locally, and not everyone in Toronto's Little Italy made a match across great distances. But normative expectations accompanied common practices. Area residents who chose to depart from local social custom were often aware they were doing so, especially in South Philadelphia, where the neighborhood demanded constant protection. After her father's death in the mid-1960s, Cathy S. began to live what she later remembered as "a wild life," going to bars and clubs in Philadelphia's Centre City. Although Cathy's mother was more "liberal" than her father had been, she remained concerned about the responses of neighbors to her daughter's unusual social activities: "She knew that everybody used to sit out . . . and if they saw me leaving all dressed up to go on the bus or something like that they knew I was going out some place." "The neighbors are going to talk about you," her mother would admonish, sparking a recurring and passionate argument: "I don't care what the neighbors say,'" Cathy would retort, provoking an escalating series of threats and counter-threats. Deviation from local customs came at a cost.[44]

Cathy's experience was not merely an exception to prove a rule; rather, she also represents the circuitous fashion in which locality could continue to matter even to those who had strayed from the neighborhood. In one of her first conversations with her future husband—whom she met at a downtown bar—they discovered that they were both from South Philadelphia. Tom, who was of Polish ancestry, revealed that he lived on the eastern outskirts of the Italian section of South Philadelphia, some six blocks from Cathy's home. Such differences of ancestry and location mattered: Cathy would joke that he needed a visa to travel along Greenwich Street into the Italian district. Still, their mutual connection to South Philadelphia sparked conversation in a crowded bar, and surely it facilitated their decision to reside, after their marriage, in the house at the heart of Annunciation Parish where Cathy had been born. Despite going to great lengths to escape the restrictive networks of the neighborhood, Cathy ended up remaining very much at home. In a different setting, perhaps Toronto's Little Italy, where a young woman's wanderlust was received on very different terms, Cathy's travels might have carried her a good deal farther.

5 Breaking the Mold: Work and Postwar Ethnicity

Vince P. was a child when his family relocated from Maierato, Calabria, to Canada. In the years that followed, his father, Paolo—like many Italian immigrants of the 1960s, a farmer turned construction worker—typically left the house early in the morning, headed, from his son's perspective, to "some faraway construction site."[1] After a long day of labor, Paolo would return home:

> Haggard, sweaty, and tired . . . He put down his black, dusty lunch box and removed his hard hat. Then, sitting in a low chair in the kitchen he proceeded to remove his mud-encrusted work boots. His feet were sweaty and smelly, the source of many an affectionate joke in the family.[2]

Years later, when the two drove through Toronto together, former sites of drudgery were transformed into landmarks of autobiography. Paolo would gesture to a building that he had helped to construct, recalling the hardships of hauling bricks and laying foundations during harsh Toronto winters, "rough supervisors," lax safety regulations, as well as the "good projects" that he had worked.[3] Personal history, for both father and son, was marked here and there, in travel about the city. In the remarkable Italian construction niche in postwar Toronto, Paolo had found work that mirrored the rest of his social experience. Construction work required Italian Torontonians to traverse long distances within a metropolitan ethnic labor niche.

Paolo lived at the heart of Little Italy, first on Grace Street and then Euclid Avenue, and traveled to work in one of the city's most complex industries. Small construction businesses sprung up for lucrative jobs, then often fell quickly out of existence. Workplace arrangements were provisional. Crews came together in peak season and for individual contracts and then disbanded when work was scarce. Workers and employers gathered at sites spread across the metropolitan area, and a completed job often spelled the end of any particular labor arrangement. New groupings of firms, workers, and unions coalesced in a new location for the next job.[4]

The construction industry was the largest and most important Italian labor niche in postwar Toronto. In the rapidly expanding city, new building abounded, and Italians flooded into demanding and dangerous construction jobs.[5] Certain construction trades, such as lathing and dry walling, were almost exclusively populated by Italian ethnics. Within such pockets, Italians were employers and employees, union bosses and rank and file.[6] Connections among conationals, and often friends and family members, animated a crucial sector of the economy. Informal social relations and formal contracts overlapped in the construction industry, and workers navigated both, as well as the metropolitan area, in the company of co-ethnics. The elasticity of Italian social ties in Toronto fit perfectly with the demands of the construction trades (figs. 34–35).

While Italians continued to concentrate in particular fields of employment in Toronto, in Italian South Philadelphia a contrasting shift was underway. Whereas previous generations of Italian Philadelphians had used ethnic social networks to negotiate labor markets, in the decades after World War II this strategy waned. Industrial era Italian labor niches faded from existence. While they maintained their insular neighborhood in South Philadelphia, Italians increasingly broke free of Italian bonds when they sought work. Thus, in Italian South Philadelphia work and social life became increasingly dissociated, while ethnic labor niches persisted in Toronto. In this final fashion, Italian ethnicity in the two locales grew increasingly apart.

Working experience plays a central role in understandings of urban ethnicity. Both of the neighborhoods under consideration here owe their origins to labor migrations. Urban Italian enclaves arose in North American cities because of ethnic networks that connected migrants at both work and home. Working life has rightfully held a central role in

FIGURE 34 Italian workers at a demolition site on Bay Street, Toronto, 1973. Photograph courtesy of Vincenzo Pietropaolo.

FIGURE 35 Construction workers in Toronto, 1973. Photograph courtesy of Vincenzo Pietropaolo.

historical explanations of the process of migration, household and family dynamics, and ethnic politics, as well as the disintegration of ethnic bonds and neighborhoods over time. Thus, the changing relations between work and social life in the postwar era, in particular in Philadelphia, signaled a profound shift in the very character of ethnicity.[7] South Philadelphians who otherwise clung to tradition broke free of previous models of working life.

The uneven distribution of ethnic groups across labor markets illuminates the importance of ethnicity to job choices. Disproportionate concentrations of members of particular ethnic groups in particular jobs arise because of the demands of the local economy, the skills and needs of immigrant populations, ethnic networks among job seekers, and preferences and prejudices among employers.[8] Previous scholars have designated ethnic niches in instances where an ethnic group's presence within a given industry is at least one-and-a-half times greater than its representation in the labor force overall.[9] While this precise threshold is somewhat arbitrary, it provides an established guide to identifying the uneven distribution of workers across jobs, and showing where ethnicity shaped occupational choices. In the case of postwar Toronto and Philadelphia, this analysis reveals the divergent role of Italian ethnicity in the two postwar labor markets. Whereas Italian ethnicity faded from significance in Philadelphia's workforce, it continued to matter to Italian job seekers in Toronto.

In Philadelphia, postwar developments diverged markedly from the past. At the close of World War II, Philadelphia's Italian population occupied a distinctive place within the metropolitan economy. With its factories still churning, Philadelphia retained many characteristics of its industrial heyday, including ethnic concentration within special niches in the economy. In 1950, Italians in the metropolitan region remained rooted in traditional labor niches. In the forty years that followed, the connections among Italian ethnicity, gender, and the search for work shifted dramatically.

In 1950, Philadelphia as a whole boasted a large and diverse regional economy.[10] Manufacturing claimed the greatest share of the metropolitan workforce, employing 36 percent of laborers. The next two most significant employment sectors were sales, with 20 percent of workers, and the service sector, which employed 18 percent.[11] No other sector included more than 10 percent of the workforce. Within each sector, Philadelphians found many ways to make their livings. The manufacture and sale of apparel and clothing accessories employed 6 percent of Philadelphians, making it the largest single industry in the city. The next most common lines of work, employment by the Federal government or by one of the hundreds of bars and restaurants dotting the city, each claimed 3 percent of workers. However, a large majority of Philadelphians, 71 percent, worked in industries that employed 2 percent of workers or less.

Italian ethnic Philadelphians were considerably more concentrated. The 1950 census does not report ethnic ancestry, but it indicates every individual's birthplace as well as the birthplaces of their parents. Hence, analysis of Italian labor patterns in this census year includes only first- and second-generation Italian immigrants. Among these, many worked within ethnic niches. Among the hundreds of industries in the city, seventeen constituted Italian niches, and Italian overrepresentation within niche industries was frequently quite considerable. In all, the niches employed 39 percent of Italian workers in the city in 1950, meaning that a goodly number of Italians found themselves in workplaces with a high number of coethnics. Among the niches, none compared in size to the manufacture and sale of apparel and clothing accessories, an important Italian labor niche in Philadelphia since the early twentieth century.[12] In 1950, the apparel industry employed 17 percent of all Italian workers. Among these, the vast majority, more than nine in ten, worked as either "operatives" or "craftspeople," in the manufacture of clothing.

In 1950, Italian men and women shared important labor niches, but work choices were also shaped by gender. Italian men specialized in hotels and lodging and the making of rubber products, niches that they did not share with Italian women. In addition, Italian men were especially concentrated in the apparel industry. Whereas the entire workforce making and selling apparel was almost two-thirds female, among Italians the split was almost even, with men comprising just less than half of Italian apparel workers. Italian men were overrepresented in the industry by a factor of 4.5, as against 3.3 for women.

Italian women, meanwhile, concentrated in some industries and jobs that drew few of their male counterparts. Women worked in clerical, sales, and especially, productive capacities in a range of industries and businesses. They concentrated in retail sales of "general merchandise," which occupied 7 percent of Italian women, and laundry services, at 4 percent. However, as with men, apparel and accessories was the most important niche for Italian women. Manufacturing in this industry alone occupied 30 percent of Italian women workers.

The most notable difference between the work habits of Italian women and Italian men was the percentage of each found within a niche industry. Among all working Italian males, 36 percent labored within an ethnic niche. The largest niche, apparel and accessories, employed 8 percent of the total Italian male labor force. By contrast, a significant majority of working Italian women, 68 percent, worked within an Italian labor niche (fig. 36). Their niche in the manufacture of clothing was far larger than any labor niche for men. Italian men and women, therefore, had different experiences in Philadelphia's early postwar labor market. For women, Italian ethnicity entailed labor force participation alongside other Italians, male and female. For men, ethnicity and labor niches were part of working experience, but not an overriding fact of working life. Italian men could and did seek employment outside of the city's Italian niches, within industries that would have included only a small number of their coethnics.[13] Such opportunities were far rarer for women.

While employment discrimination may have contributed to the Italian labor niches, Italians who worked in the city in the years after the war instead emphasize the social networks that led to their choices of work. Labor within the community was especially typical for the youngest workers in South Philadelphia. Anthony S., born in 1931, held a variety of jobs in the 1940s, each of which kept him firmly entrenched in local Italian social networks. Whether shining shoes in local pool halls, delivering the *Philadelphia Inquirer* on neighborhood streets, or

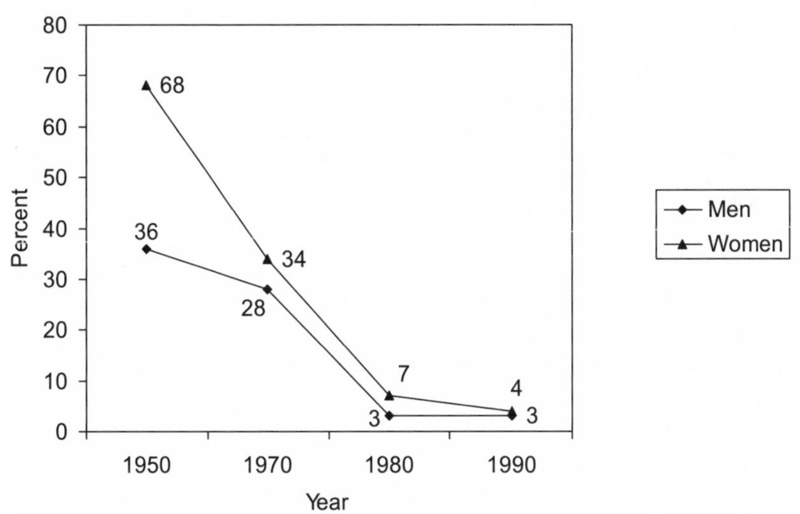

FIGURE 36 Percentage of Italians in ethnic labor niches by gender, Philadelphia, 1950–1990. SOURCE: *Integrated Public Use Microdata Series: Version 4.0,* Minnesota Population Center.

selling fresh baked pretzels from his red wagon on Saturday mornings, Anthony remained close to home when he went to work.[14] If children epitomized labor within the ethnic fold, such connections continued well into adulthood. Cynthia C. explains that when she began working in the banking industry in the 1950s, family contacts helped her to secure a job. "In those days," she recalls, Philadelphia banks were eager to offer jobs to relatives of existing employees: "In my case, the Philadelphia Saving Fund, I had two uncles and three or four cousins working there." While the choice to work among family was natural for Cynthia, the local pastor pushed her in other direction: "[he] wanted me to go interview at Beneficial Savings Bank, which was known as the Catholic bank."[15] Thus, the overlap of work and social life, while useful, could also generate tension. As Cynthia navigated the workforce, competing social influences—of family connections and the parish priest—shaped her expectations and choices. In the early postwar period, young Italian women were required to navigate the overlap between social and work life while taking advantage of the opportunities it presented.

During the decades that followed, Italian labor niches collapsed, especially in the experiences of working women. With the economy moving away from manufacturing, the traditional connection between Italian ethnicity and work disappeared. Cynthia's experience notwithstanding, Italians never established concentrations in services or sales that compared to their manufacturing niches. By the last decades of the century, Italians were evenly spread across the economy. Italian ethnicity had ceased to shape working life.

The workers recorded in the data from 1970 provide crucial evidence of the shift out of niche industries. With the census remaining silent on the question of ethnic ancestry, the Italian population captured in the census still included only first- and second-generation Italian Americans. By 1970, this group scarcely represented the entire Italian ethnic population of the city. Almost two generations removed from the decline of Italian immigration to the city, the first- and second-generation immigrants were the oldest Italian workers in the city. In 1970, the mean age of workers with Italian parents or who were themselves born in Italy was forty-seven years, as against thirty-two years of age for the entire workforce.[16]

Nonetheless, the Italian workers of 1970 had made a significant move out of niche industries. The large niche in apparel had fallen precipitously, now occupying only 10 percent of Italian laborers. Construction emerged as an important niche, employing 8 percent of Italians, but overall the percentage of workers in niche industries had fallen to just over 25 percent of the Italian labor force, a decline of almost one-third. Italian women saw an especially high exodus from niche labor. Whereas in 1950 more than two-thirds of Italian women worked in niche industries, by 1970 only one in three did so. Women were still more likely than men to work within an Italian niche, but now, as for men, niche labor was a minority experience—most Italian women worked in an industry wherein the Italian population was not especially prominent.

Italian labor niches rapidly declined as industrial labor faded from prominence in the metropolitan economy. Those captured in the 1970 sample were the most likely to be caught in declining industries and the least likely to be able to escape a niche for other gainful work. In addition, older workers were the least likely to deviate from earlier gendered patterns of labor. Yet, in 1970, many had done just this—an older generation of workers anticipated the trends that would be visible within a wider swath of the Italian ethnic population in coming years. They abandoned niche industries and relaxed the special imperative for Italian women to work with coethnics.

South Philadelphians describe workplaces of the 1960s and 1970s as sites of integration. In the mid-1960s, when Cathy S. worked in the billing department of a lighting store in the northeastern section of the city, few of her coworkers were Italian. Her closest friend at work, Mary-Anne, came from an Irish neighborhood in the northeast, Juniata. For Cathy's mother, herself South Philadelphia born and raised, Mary-Anne represented "the other end of the city, the other end of the world."[17] Cynthia C., who initially secured a job through family contacts, found herself in changing surroundings as she moved up the corporate ladder. When she assumed a leading role in the mortgage operations of the Philadelphia Savings Fund, her coworkers at the head office came from varied origins and lived in neighborhoods throughout the metropolitan region.[18]

By 1980, Philadelphia's postwar economy had been thoroughly reshaped. Manufacturing had fallen from its position as the largest employment sector. Services now ranked first with 30 percent of the workforce, and manufacturing and sales ranked second and third, with 23 percent and 21 percent of workers, respectively. Among the individual industries, educational services now placed first with 9 percent of the workforce, followed by hospital work at 5 percent. The metropolitan economy was a different place than it had been decades prior, and Italians had adjusted. In the 1970s and 1980s, Italians in metropolitan Philadelphia all but completed a move out of niche labor, settling into an even distribution across the changing economic landscape of the metropolitan area.

With little ongoing Italian immigration to Philadelphia, the character of the Italian workforce shifted over time. In 1980, the census asked individuals to report their ancestry; most workers who described themselves as Italian on these grounds were generations removed from immigration. While many among them continued to live in Italian neighborhoods, others shared their most important social ties with non-Italians.[19] Like their peers in the rest of the country, many of these Italians inhabited ethnically diverse neighborhoods, churches, and families.

Integration was also evident in the workforce. In 1980, Italian ethnics in Philadelphia distributed their work in accordance with citywide labor patterns. Niches remained in a number of industries, but together they employed only 5 percent of the Italian workforce. Italian women concentrated only slightly more than men, with 7 percent working in niches. With services and sales further outpacing manufacturing within the metropolitan economy by 1990, older niches such as the manufacture of clothing all but disappeared. Italians established a small niche in "miscellaneous personal services," but the vagueness of this census category makes any conclusion from this concentration difficult. In all, only 2 percent of Italians workers and 4 percent of working Italian women remained in niche industries. The era of Italian niche labor had ended—Italians now distributed themselves evenly across the metropolitan economy. Ethnicity, and its once powerful combination with gender, had faded from importance in the distribution of Italian labor across the metropolitan economy.

Although part of the decline in Italian labor niches can be attributed to census definitional changes, the data nonetheless indicate that ethnicity was actually becoming less important to workplace decisions. By capturing Italian ethnics beyond the second generation in the 1980 and 1990 censuses, the data are skewed toward showing greater ethnic integration over time. However, the 1970 data hazard against interpreting the decline of Italian labor niches as a mere artifact of definitional change. In 1970, first- and second-generation Italian Americans had already begun the exodus from niche labor. Their distribution across the labor force in 1970 suggests an important trend afoot: the shifting economy offered little opportunity for Italian niche labor. The industrial economy had come of age in the era of mass Italian immigration to Philadelphia at the turn of the twentieth century. Immigrant networks had facilitated connections between workers and employers as Philadelphia's factories were built, and the Italian niches established in the immigration era persisted into the early postwar period. However, as old industrial era niches declined and the service-oriented economy took shape, Italian ethnicity ceased to play this role. The postwar economy played host to niches comprised of new immigrant groups, but among Italians connections between ethnicity and work grew less common.[20] By 1970, immigrants themselves and their children had begun to adjust to the new economy as they distributed themselves more evenly across the workforce.

The changing economy brought a decline in Italian ethnic niches, but their near disappearance also required a generational shift. Integrated into the wider society surrounding them, the new generation of Italians completed the exodus from ethnic niches. When they went to work, their Italian roots hardly figured in their labor choices.

In Toronto, by contrast, Italian labor niches persisted in the postwar era. As Italian immigrants poured into the city, and in the decades thereafter, Italian ethnic origins continued to play a pivotal role in working life. As in Philadelphia, Toronto's early postwar economy offered diverse employment. In 1951, manufacturers employed 36 percent of all workers in the metropolitan area. The service sector was next, with 21 percent of the workforce, and trade third, with 19 percent of all workers. The construction industry, the most important Italian niche in postwar Toronto, employed 7 percent of metropolitan Toronto's workforce.[21]

For the most part, Italian men and women occupied separate niches

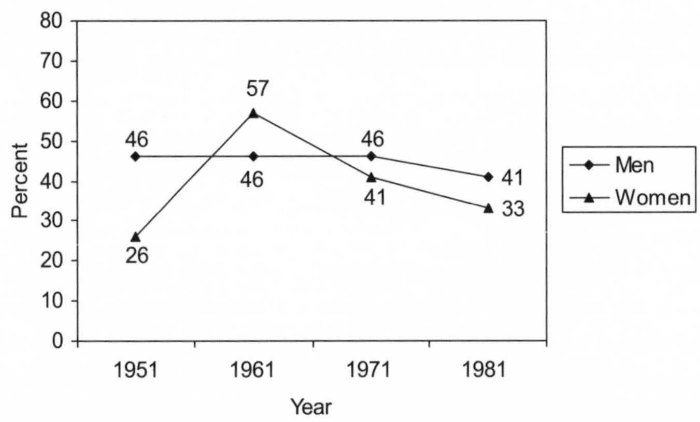

FIGURE 37 Percentage of Italians in ethnic labor niches by gender, Toronto, 1951–1991. SOURCE: Franca Iacovetta, *Such Hardworking People,* appendix, tables 11–14; Census Special Tabulations for Author, 2007.

in early postwar Toronto. The niche in construction was almost entirely comprised of men. Their presence in building trades was 3.6 times greater than their representation in the workforce. A remarkable 33 percent of working Italian men found jobs in the industry, making it the largest Italian labor niche for either sex. A distant second among male niches, grocery stores employed 6 percent of Italian men. Other, small areas of niche employment included barbershops, agricultural work, and the manufacture of leather and tobacco (fig. 37).

Italian women in Toronto found different areas of niche employment. In 1951, 18 percent of Italian women made clothing, their presence within the industry was 2.7 times greater than their representation in the workforce overall. Italian women were even more concentrated in laundry services. With an Italian presence 4.4 times greater than that in the entire workforce, the small laundry niche was the city's densest concentration of Italian female labor. Italian men and women thus occupied largely distinct labor niches in Toronto. Italian men were slightly underrepresented in the making of clothing, and few Italian women found employment in construction. Gender division set Toronto's Italian laborers apart from those of early postwar Philadelphia, where Italian men and women shared specialized industries.

In all, a fairly large percentage, 37 percent, of Toronto's Italians worked in ethnic labor niches in 1951, but the role of gender in this choice was the inverse of that in Philadelphia. Italian men in Toronto were more likely to be found in niche labor than females. Whereas 46 percent of Italian men worked within a labor niche, only 26 percent of Italian women did the same. Thus, whereas women in early postwar Philadelphia were especially funneled, or drawn, to work with coethnics, in Toronto it was initially men who concentrated among other Italians.

The sex difference in Italian niche labor in the early postwar years reflected the gender dynamics of Italian migration to Toronto during this period. Between 1946 and 1955, some 60 percent of Italian immigrants to Toronto were men, many of them having been admitted as laborers.[22] Immigrant men who flowed into Toronto on "bulk orders" by employers and other formal contracts naturally entered niches, since employers brought many at a time. Others left contracts elsewhere in the country and wound their way to Toronto where social networks steered them towards niche labor.[23] Italian women, many of whom had arrived to the city decades prior, were more likely to work outside of ethnic niches.

As immigrants continued to pour into the city, the sex ratio grew more even and the particular propensity of males to seek ethnic labor niches disappeared. By 1961, as in Philadelphia, Italian women were more likely to work in ethnic niches than men. This shift was not due to a decline in male niche labor. In 1961, like a decade prior, 46 percent of Italian men in Toronto worked within an ethnic niche, among these 8 in 10 worked in construction. Women's participation in labor niches, however, had dramatically increased: 57 percent labored within an Italian niche. The concentration of women in the manufacture of clothing had increased slightly to 19 percent of the female Italian workforce, and it was now joined by other important clusters. Italian women concentrated in the making of leather, with 4 percent of Italian women workers, textiles, 7 percent, and food and beverages, 6 percent. At the same time, Italian women had developed a niche in the emergent service sector; 16 percent of Italian women worked in miscellaneous personal services, where their representation was almost twice that in the labor force overall.

The jump in Italian women's niche participation reflected the presence of new immigrant women. Newly arrived Italian women found work alongside other Italians. One woman who immigrated to Toronto in 1958 recalled walking miles along Spadina Avenue, from Davenport to Queen Street, in search of work:

> I was walking, just going from one place to another, and the only things that I knew to say in English was, "Did you have work for me?" That my brother taught me. And I went in, in many places . . . And they look at me probably, I don't know what they would say, but probably they ask me, "Do you have

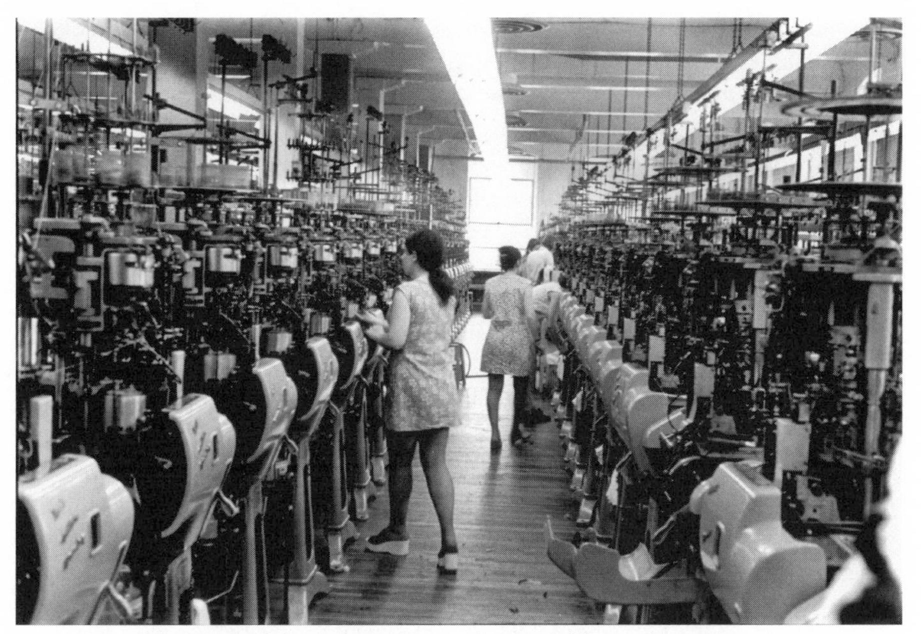

FIGURE 38 Italian garment workers in Toronto. Photograph courtesy of Vincenzo Pietropaolo.

> any experience?" And I didn't understand what they were saying so I was just waiting for them to say "Yes" or "No."[24]

Finally, she found a blouse manufacturer who already employed other Italian women, one of whom offered to translate. The next morning she began work, becoming yet another Italian woman in the large niche in apparel manufacture (fig. 38). On the face of it, her story has an apocryphal air. How could someone speaking only one sentence of English seek a job by foot in the city? Nonetheless, the story reflects a fact of working life in early postwar Toronto—Italian women (and, to a lesser extent, men) found work beside coethnics and fellow immigrants. A coworker who was a fellow immigrant, especially one who had spent a few additional years in Canada, could be a considerable asset.

Italian workers benefited from the presence of coethnics, but employers also prized Italians, in particular, for certain jobs. Classified job advertisements in the *Corriere Canadese,* the city's leading Italian-language newspaper, often specified, in Italian, that employers sought Italian laborers. In 1960, Vito Giovannetti sought an experienced Italian cement worker as well as an "Italian engineer" who spoke fluent English.[25] A year later a tailor sought a "ragazzo Italiano" to work as his assistant, and a laundry promised work for a "ragazza" who spoke Italian as well as English.[26] A plumber, recognizing the particular demand for service that could be delivered in Italian, advertised himself as a "plumber Italiano."[27] Sensing a rapidly expanding new market for its products, Avon sought "signore Italiane" or "madri di famiglia," who could earn $10 a day selling cosmetics part-time in their own neighborhoods.[28]

If Italian women sometimes preferred niche employment, they also lacked other opportunities. In some instances, immigrants recalled explicit prejudice on the part of employers. One Italian immigrant to Toronto described rampant discrimination against Italians stemming from the Italian participation in World War II: "If you went for a job . . . instead of saying, 'What can you do,' they say, 'What nationality are you,' you say 'Italian'—'No job for you.'"[29] In the decades that followed, the jobs available to Italians, and Italian women in particular, continued to be constrained by employers, who often discriminated on the grounds of both ethnicity and sex.[30] Many immigrants funneled into Italian niches would likely have preferred other alternatives. Paul N. remembered that when his family arrived to Toronto in the late 1950s, his mother "went to work in a factory . . . making coats, pants, shirt and things like that . . . that's what most of the women did in those days." For his mother, however, the choice entailed hardship: "[M]y mother had a store all her life [in Italy]. So I think it was most difficult for her when she came here because all of a sudden she had to go work for somebody else."[31] Niche labor reflected barriers as well as opportunities.

The particular propensity of Italian women to find niche labor in 1961 also stemmed from constraints imposed from within the immigrant community. Francesca S., who arrived to the city in the 1960s, recalls that Italian ideals conflicted with Canadian realities as young Italian women entered an urban industrial workplace. Social norms continued to valorize sheltered, domestic young women, but economic imperatives demanded their workforce participation. This contradiction was resolved, at least in part, by labor niches. Francesca recalls, "before the girl . . . came to Toronto a *paisana* or a godmother or an aunt or a cousin would . . . make sure that this girl . . . would find a job where she worked." The guardian "*paisana,*" usually an older, married woman, "would become a surrogate mother, would ensure that this girl was under her supervision."[32] Italian labor niches preserved a semblance of traditional Italian community, and in particular its gender norms, even in the midst of industrial production.

Over time, such constraints relaxed. As in Philadelphia, women's niche labor in Toronto declined as they moved out of manufacturing. By the 1970s, Italian women in Toronto were less likely to work in niche industries than their male counterparts. By 1981, only 33 percent of Italian women worked in a labor niches, representing a decline of over 40 percent in two decades. In contrast to Philadelphia, however, a notable minority of women continued to work in manufacturing niches. In 1971, 8 percent of Italian women made clothing and 41 percent engaged in manufacturing overall. In 1981, the manufacturing sector employed more than 30 percent of Italian women, who concentrated most heavily in knitting mills and in the manufacture of furniture, leather products, and clothing. Still, the majority of Italian women had moved out of manufacturing and, at the same time, out of Italian niches.

Although the distribution of women across sectors differed in the two cities—with Italian women in Toronto more likely to remain in manufacturing—women's labor niches played a similar role in each economy. In both cities, Italian female niches were important to manufacturing, but not to other lines of work. In Philadelphia, Italian women remained within labor niches in the early postwar period, as industrial production in the city remained high. As both men and women moved out of manufacturing in Philadelphia, Italian women's participation in niches declined to that of men. In Toronto, niche work persisted to the extent that women continued to work in manufacturing. In other sectors, as in Philadelphia, women worked alongside non-Italians. In both cities, service occupations, many of which were widely understood as suited to female workers, posed a lesser threat to the expectations generated by gender and ethnicity. Female Italian labor niches were an artifact of industrial labor.

For Italian men in Toronto, labor niches remained stable despite changes to the economy, largely because of the enduring importance of construction. In 1981 23 percent of Italian men continued to work in the remarkable niche in construction, down from 28 percent in 1971. Unlike women, men also developed enduring niches in the new economy. In both 1971 and 1981, retail and personal services—more specifically grocery stores and barber shops—employed more Italian men than any single category of industrial manufacturer.

Even in industries in which Italians did not predominate, Italian social ties could be important to securing employment. Paul N. began a long career in movie theaters as an usher at the Studio Theatre, which was located near his home in Little Italy. The theater, which showed Italian films, was a frequent haunt of Paul's and his friends before he was hired. When he decided "I'm tired of paying," Paul approached the manager—like him an immigrant from Puglia—and inquired about a job. As the two became close friends, Paul's responsibilities increased, and he ultimately rose from usher to the manager of several large theaters.[33]

Paul's work experiences also suggest that labor niches could foster social ties among members of the residentially dispersed Italian community. Paul met his wife, who resided far from Little Italy in the northwestern section of the city, while he was on the job. During a brief hiatus from his work in movie theaters, Paul found summer employment with a plastic company where he was "the coffee boy for the vice president." Ordered one day to wash his boss's Cadillac, Paul thought instead to solicit a volunteer from the factory: "So I went to the factory, I see this girl and I say, 'Would you wash my car' . . . She said, 'No'!" Intrigued by the spirited factory girl, Paul returned at the end of the day to offer her a ride home in his boss's car, pretending it was his own: "And that's how it started." In the years that followed, his wife would maintain that "she fell in love with the Cadillac."[34]

Italian labor niches in Toronto persisted in the decades after World War II. At a time when Italians in Philadelphia integrated into the wider workforce, more than 40 percent of Italian men and more than 30 percent of Italian women in Toronto continued to work in niche industries. The largest (and most noted) concentration was in construction, but Italians also worked with coethnics in the manufacturing and service sectors. Even as a majority worked outside of these specialized ethnic industries, niche labor remained a key dimension of workforce experience for Italian Torontonians.

But what of Italian South Philadelphia and Toronto's Little Italy? Thus far, conclusions have been drawn from data on a metropolitan scale; offering only limited insight into the specific neighborhoods under consideration here. However, the limited data on industrial occupation by census tracts offer some suggestion that citywide patterns also characterized life within the Italian neighborhoods. In Toronto, tract-level data on workforce participation fail to divide the local population by ethnicity, so its usefulness is compromised by the diversity of Little Italy. In 1971, even after the heavy influx of Italian immigrants into the city, Italian ethnics comprised only a third of residents inside the parish boundaries of St. Francis.[35] Thus, data from the tracts in Little Italy capture the work patterns of many non-Italians. Nonetheless, tract-level data suggest that workforce experience in Little Italy conformed to wider metropolitan trends. Men residing in the parish were heavily employed in construction throughout the postwar period. In

1981, 20 percent of men in the area worked in construction, a number close to the citywide figure for Italian men. If non-Italians in the district were less likely to find employment in this field, then perhaps the Italians in Little Italy were especially likely to work within the construction niche. Male residents of the tracts were also concentrated in the citywide Italian niche in food services, which, along with accommodations, employed some 10 percent of men in the area. Like Italian women in the rest of the city, local women were overrepresented in manufacturing, which employed 30 percent of females in the tracts. Thus, while the tract data are not specific to local Italians, it seems reasonable to conclude that citywide trends were replicated in Little Italy. The same Italian niches persisted from the 1960s to the 1980s.

The Italian parishioners of Annunciation and St. Thomas Parishes were heavily concentrated in three Philadelphia census tracts where they comprised some 60 percent of residents for much of the postwar era.[36] Data from these tracts, therefore, reveal a good deal about labor patterns among the Italian Americans at the center of this study. Unfortunately, the data from the tracts are poor, reporting the labor patterns for both sexes together and using very broad categories to describe industrial concentrations. The data are most revealing for 1960. In that year, 27 percent of workers in the two parishes were engaged in the manufacture of textile and apparel; their presence in the industry was 2.3 times greater than in the labor force overall and even greater than the concentration of citywide Italians in the niche.[37]

The tract level data from 1970 onward use very large industrial categories, ceasing, for example, to separate textiles and apparels from other forms of manufacturing. Given these larger industrial categories, it is impossible to identify niche concentrations by South Philadelphians after 1960. The only information available—the data for Italians in the metropolitan area—shows a decline of niche labor. South Philadelphians, being especially concentrated in apparel in 1960, may have been slower to move out of niches than other Italians in the metropolitan area, but no data are available to support this supposition.

Italian South Philadelphia and Toronto's Little Italy come more clearly into focus in another source: tract-level data on travel to work. Journey to work data suggest that Italian South Philadelphians, like their coethnics elsewhere in Philadelphia, did indeed break the mold of local life to pursue jobs in an integrated labor force. While journey to work data do not indicate the particular jobs of Italian South Philadelphians, they do indicate (sometimes indirectly) where they went when they set out for their jobs every morning. Journey to work data

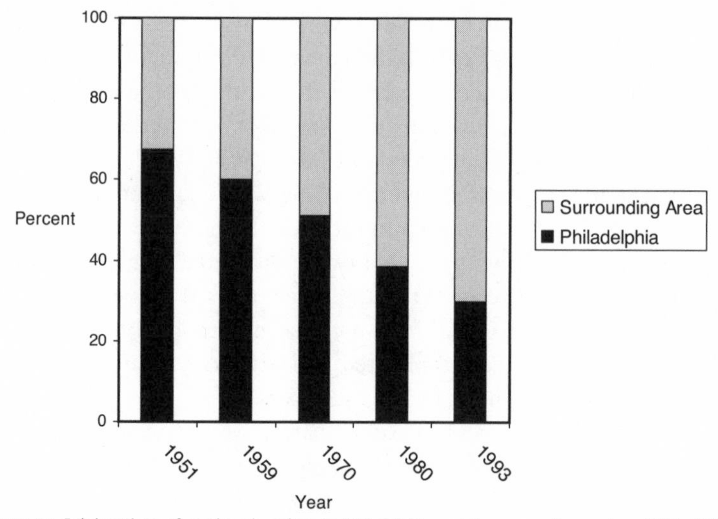

FIGURE 39 Job locations of employed workers in Philadelphia and its surrounding region, 1951–1993. SOURCE: *Philadelphia: Neighborhoods, Division and Conflict in a Postindustrial City* (17); *County Business Patterns, 1993: New Jersey* (CBP-93-32); *County Business Patterns, 1993: Pennsylvania* (CBP-93-40; Washington, DC: U.S. GPO).

demonstrate that Italian South Philadelphians moved well beyond their neighborhood when they set out to work. The local bonds that characterized the rest of social experience did not hold sway in the search for employment.

As in the rest of social experience, the geography of work reflected wider urban contexts. In Philadelphia, the decentralization of labor force opportunities meant that jobs were dispersed across an ever greater area. In the postwar era, the city of Philadelphia, comprising 143 square miles, lost its position as the center of employment within the larger metropolitan area. Each decade after World War II saw jobs further dispersed into the seven counties surrounding the city, which together sprawled across more than 3,400 square miles. In 1951, as Philadelphia neared the end of its industrial heyday, the city still claimed more than two-thirds of employment within the region. By the 1990s, this figure had reversed, with two-thirds of jobs located outside of the city of Philadelphia (fig. 39).[38] When South Philadelphians went to work, they followed the center of gravity within the region.

Even in the early postwar era, South Philadelphians traveled to jobs distant from their homes. City planning studies in the 1950s and 1960s describe workers flowing out of, as well as into, South Philadelphia in the course of their daily commutes. Indeed, the labor flows connecting South Philadelphians to the rest of the city roused some degree of concern about the dearth of public transportation available in the area. A report in 1959 worried that employees working at the piers, oil refineries, and naval installations in South Philadelphia lacked adequate means of reaching those areas, while tens of thousands of South Philadelphians commuting elsewhere for work were delayed on the congested streets carrying them north to the city center and beyond.[39] A study in the mid-1960s reported similar findings. Analysis of traffic flows along Chestnut Street in the central business district found South Philadelphians headed out of the area to places of work throughout the city and beyond.[40] South Philadelphians left the area, if often slowly, in their travels to work.

Census data from the same period confirm the findings of the traffic surveys. In 1960 and 1970, the census reported the means, but not duration, of workers' daily travel to work. The great majority of South Philadelphians, like workers in the rest of the city, used automobiles or public transportation to get to their jobs. In 1960, 38 percent of the workers in Annunciation and St. Thomas Parish drove to work and 50 percent took public transportation, numbers similar to those for the entire city. Only 11 percent of workers in the two parishes walked to work, a figure almost identical to the citywide number. Ten years later, travel to work had hardly changed. Workers now split evenly between cars and public transportation, each with 45 percent of workers, and the remaining 10 percent walked. If walking to work was a barometer of working close to home, few workers in Italian South Philadelphia did so; they were no more likely to live within walking distance of their work than other Philadelphians.

In 1980 and 1990, an additional census variable, duration of travel to work, allows more precise analysis. The modes of transportation to work in 1980 in both South Philadelphia and the wider city resembled figures from earlier censuses. In 1980, 47 percent of workers from Annunciation and St. Thomas drove cars to work, and 37 percent rode public transportation. The percentage walking to work, 14 percent, was now slightly higher than in the past. Most traveled a good distance from home. Very few residents of Annunciation and St. Thomas Parish reported travel times short enough to keep them within the boundaries of Italian South Philadelphia.[41] A large majority, 82 percent, traveled for 15 minutes or more to work. Many reported trips a good deal longer; 51 percent traveled for 20 to 44 minutes and another 15 percent traveled for 45 minutes or more. With most riding in cars or using public transit, trips of 15 minutes or more would have carried most workers outside the small radius of Italian South Philadelphia. Ten years later, the story was virtually unchanged. The majority of South Philadelphians traveled to work by car or public transportation and reported lengthy trips. South Philadelphians broke free from local boundaries as they traveled to work.

Unfortunately, the aggregate data, which provides information specific to South Philadelphia, fails to divide the journey to work by sex, so a consideration of the connections between ethnicity, space, and gender in the labor force is limited to metropolitan data. In 1970, Italian women in the metropolitan region were especially likely to work close to their homes. In that year, 17 percent of Italian women walked to work, as against 6 percent of Italian men and 10 percent of non-Italian women.[42] For Italian women, then, some of the localism of ethnic life found its way into working experience. Italian women workers in 1970 were a good deal more likely to work within the narrow geographic boundaries of the rest of their social lives. Still, local work was a minority experience even for Italian women—the great majority used cars and public transportation to get to work.

Most likely, the expectation that Italian women would work locally was already on the wane by 1970, along with the special expectation that they work in ethnic niche industries. By 1980 and 1990, Italian women were no longer especially likely to walk to work. All women in the metropolitan area reported shorter travel times than men in 1980 and 1990, but Italian women showed no particular tendency to work close to home. If the responsibility to care for children or concerns about distant travel through the city kept women workers closer to home, this was not an especially Italian phenomenon.[43]

In the early postwar era, Italian women in Philadelphia worked in accordance with restrictions associated with their sex and their ethnic origins. Much more than their male counterparts in these first decades after the war, their labor experience was shaped by the presence of fellow ethnics in their industries. Much more than men of Italian origins or women of other backgrounds, they worked close to home. By 1970, such restrictions were on the wane. Italian women moved out of niche industries into the wider workforce, and most likely, they traveled further to work. By the 1980 and 1990 censuses, only a small minority of Italian women worked locally or within niches, the great majority spread across the industrial and geographic landscape like other workers. The shifting characteristics of women's work changed what it meant to be an Italian woman. As new postwar generations came of age, being an Italian woman was no longer a status to which special work restrictions were attached. Instead, an Italian woman of the late century could expect full participation in the metropolitan workforce, at least in terms of where she worked and with whom.

In the postwar era, Italian South Philadelphians seldom worked locally. Although a small proportion traveled only a short distance to work, the prevailing experience was one of longer and more distant travel. Most likely, distances increased as decades passed, the city's industrial economy faded, and jobs within the metropolitan area dispersed further from Italian South Philadelphia. By the 1970s and 1980s, South Philadelphians traveled to work like their peers in the rest of the city. Although in other aspects of social life ethnicity guarded urban territory, workers moved widely across the city in their prolonged daily commutes.

In Toronto, the workers from Little Italy confronted a very different field of urban employment. Despite significant development in the wider region, Metro Toronto remained the local center of gravity. In 1960 Metro claimed a near monopoly in the region, with more than nine of ten jobs. The following decades saw significant growth in the municipalities surrounding Metro, and varied employment opportunities arose in outer suburbs. Nonetheless, Metro still claimed a large majority of jobs. In 1991, 64 percent of jobs were still located within Metro.[44] Many of these jobs remained in the city of Toronto itself, offering opportunities that would have been apparent to residents of Little Italy, who lived just to the west of one of the major nodes of employment in the region.[45] In 1971 census tracts on the eastern edge of the parish drew over 230,000 workers, comprising some 23 percent of all jobs in the metropolitan region. In 1981 more than 285,000 workers found jobs in the same tracts, or 20 percent of the employed labor force.[46] Little Italy remained an attractive immigrant destination in part because of its proximity to important job sites. Rather than using their citywide ethnic networks to find jobs, many residents of Little Italy turned to the nearby employment hub.

As a result of the wider geography of the labor market, workers from Toronto's Little Italy were significantly more likely to find local work than Italian South Philadelphians at the same dates. In 1971, 11 percent found work with the parish itself, and an additional 36 percent worked in the high employment area just to the east of the parish. In total, almost half of the employed labor force in the parish found work within just a few miles of their homes.[47] In 1981, when more than half of South Philadelphians spent 20 to 44 minutes commuting to work, residents of Little Italy stayed local. Within the employed labor force of Little Italy, 12 percent worked in the parish, with an additional 37 percent in the nearby employment hub. Italian labor patterns greatly resembled those of other Torontonians, who also came to the central area for work. Thus the spatial dynamics visible in the rest of social life were reversed when it came to work. Italian South Philadelphians, otherwise entrenched in neighborhood, found work throughout the city. Torontonians, who traveled widely in their social lives, worked close to home.

Gender played an important role in determining how far people traveled to work, with women from Little Italy especially likely to work close to their homes.[48] The dense cluster of jobs in and near St. Francis claimed more than 60 percent of women in 1971, with the percentage slipping very little in the following decade. Men were significantly less likely to work in the same geographic area. In 1971, 36 percent found work within the same local tracts that encompassed a majority of women; in 1981, 43 percent did so. The remaining male workers spread across Toronto. The northwestern section of the city, and the periphery beyond, claimed 12 percent of the male workers from Little Italy in 1971 and some 15 percent in 1981, while the others dispersed across the cityscape, finding work without a significant degree of geographic concentration. The constrained geography of female labor accords with broader trends in the labor market. Rather than reflecting an especially Italian experience, women's decisions to work close to home reflected a combination of domestic responsibilities and economic subordination felt by female laborers of many backgrounds.[49] Thus, by the 1970s and 1980s, Italians men and women in both cities traveled to work much like other workers. For Italian South Philadelphians that meant breaking out of local networks, for the residents of Little Italy it meant remaining close to home in a local cluster of employment.

The construction industry in postwar Toronto would have exemplified working life in both Little Italy and Italian South Philadelphia at any time prior to World War II. Like other immigrants at the turn of the twentieth century, Italians lived and worked together, and these two aspects of social life were deeply interwoven. Whether sojourning laborers or permanent settlers, immigrants in the industrial heyday concentrated in urban neighborhoods in large part because of their connections at work.[50]

The postwar era brought changes for Italians in both cities. In Toronto, Italian niches persisted in the labor market, even as Italian residential concentration declined. Immigrant social networks were perhaps especially well suited to the construction industry, but Italians in other walks of life continued to see the benefits of working among their countrymen. Although they moved widely about the city in the rest of social life, the labor market proved more constraining. In Philadelphia, by contrast, Italians who stuck to their neighborhood broke free of labor niches. English-speaking and resident in Philadelphia for generations, the Italians of South Philadelphia took advantage of the opportunities available in the service sector, integrating into every area of the new economy. To do so, they travelled ever further from home, abandoning the local circles that constrained them in the rest of their social lives.[51] The link between home and work—so integral to the formation of Italian neighborhoods during the industrial era—was broken, albeit in different fashions, in both cities.

Conclusion

Stephen Muzzatti and his friends drove to Woodbridge, a suburban center of Italian settlement to the north of Toronto, in search of moneyed, powerful, and somehow deeply Italian Italians. The boys, at age seventeen, had come to understand that Italian Toronto extended well beyond their own working-class district in the northwest of Toronto. Beyond Metro's limits, they believed that they would find another species of Italian Canadian. When they arrived, the appearance of Woodbridge, which they had imagined to resemble Italy itself, startled them: "There was so much space—too much space—not like farmland—but instead like a city lot suffering from gigantism."

Woodbridge, they concluded, did not look like Italy. But more than anything, Muzzatti later recalled the distance that he felt from the young people he met there. The teenagers in Woodbridge "were supposed to be like us. They were Canadian-born children of postwar Italian immigrants." Instead, differences seemed overwhelming:

> We liked heavy metal and hard core—feeding our addiction with mixed tapes. They liked something called danse (spelled with an "S"), and had a lot of records. Our uniform consisted of jeans and tee shirts—come cold weather—add a jean jacket . . . They seemed to be seasonally outfitted by

> Le Chateau. They got together at places like O'Tooles and went to dances at La Vedette Banquet Hall. We sat on lawn chairs in [the] Country Style Donuts' parking lot and talked about racing, or drank at Nick's Underground—and talked about racing . . . Their fathers were "contractors" or something called "developers"—not construction workers—and many of their mums didn't even work outside the home.[1]

These recollections suggest the limits, but also the strengths, of the foregoing analysis. My comparison of Toronto and Philadelphia has privileged prevailing dynamics of social life, identifying differences between broad patterns of activity in two separate historical settings. This approach can obscure the internal diversity within each locale. Muzzatti's story provides a corrective reminder that Italians traveling in and around Toronto could be surprised by what, and who, they found. Further, the lines of division that he describes are of scholarly interest: consumerism, class, and gender interweave with national origins in the making of individual and group identities. In Toronto, as in Philadelphia, Italian ethnics have been, and continue to be, diverse.

And yet, Muzzatti and his friends did travel to Woodbridge. They piled into a 1978 beige Toyota Tercel—"painstakingly restored to its gleaming banality through a series of all-night amateur mechanic-ing sessions"—and set out across the city. Even as Woodbridge shocked them—signaling the social distance between city neighborhoods and outer suburbs—the young men participated in the movement of Italian Torontonians about the metropolitan region. If the result was not always immediate social connection, such travel nonetheless reinforced the distinctive choreography of ethnic life in Toronto.

In this regard—in the patterned use of city space—residents of the two oldest Italian neighborhoods in Toronto and Philadelphia set ethnicity on separate courses. Italian social ties extended outward from Little Italy across Toronto in the second half of the twentieth century, while those in South Philadelphia remained rooted in locality. Tremendous increases in property values encouraged frequent housing exchanges in Toronto's Little Italy, undermining stable residential concentration. At the same time, geographically expansive networks bound Italian Torontonians together in religious and social practices. In South Philadelphia, by contrast, Italian ethnic spatial relations took shape in a context of racial and economic divide. Stagnant property values encouraged residential stability, and social and institutional life guarded urban territory. Italian ethnicity in Toronto operated on a metropolitan scale, while in Philadelphia it marked local boundaries.

These observations carry ramifications for the history of ethnic life in North America. Historians have long viewed ethnic experience as a negotiation between immigrant traditions and skills and the demands made, opportunities offered, and limitations imposed within their places of settlement.[2] Scholars who have undertaken explicit comparisons of a single ethnic group in multiple locations have regularly reported that local contexts shape ethnic experience. Samuel L. Baily's influential study of Italian migrants to Buenos Aires and New York at the turn of the twentieth century demonstrates that the experiences of Italian immigrants in the two locations differed because each city offered its own "structurally defined stage" for newcomers.[3] Baily's quantitative analysis vividly demonstrates that economic, political, and social conditions where the immigrants settled shaped the ability and propensity of Italians to advance economically, develop ethnic institutions, integrate with their non-Italian neighbors, and settle permanently in their new homes. My study reinforces the findings of other comparative historians, and in particular, Baily's suggestion that attention to the systematic patterns of social life in diverse contexts can yield specific insight into the interplay of local environments and ethnic community.

However, the evidence presented here also differs in important ways from previous research. Analysis of the ways in which ethnicity was practiced—rather than the extent to which it persisted—allows comparative research to contribute to the broader conceptualization of ethnicity. Comparison of Toronto and Philadelphia has revealed that ethnicity itself was a different kind of social bond in different contexts. This study thus makes an empirical contribution to recent theorizations of ethnicity that emphasize the ongoing activity necessary to make ethnicity an important facet of social experience. As Kathleen Neils Conzen and her colleagues argue in their seminal article, "The Invention of Ethnicity," ethnicity involves a "process of construction or invention which incorporates, adapts, and amplifies preexisting communal solidarities, cultural attributes, and historical memories."[4] Rogers Brubaker pushes such formulations even further. Drawing on E. P. Thompson's famous analysis of class, Brubaker eschews terms such as "membership" that imply a group with enduring reality. Instead, he describes ethnicity "as an *event,* as something that 'happens.'"[5] Italian ethnicity is thus not something to which a person belongs but rather something that he/she does along with other people.

While a language of "events" fits awkwardly with ongoing social processes, Brubaker's position focuses attention on pivotal aspects of ethnic experience. Brubaker stresses that his approach "keeps us analytically attuned to the possibility that groupness may *not* happen."[6] Similarly, if ethnicity is an event, then it can happen *differently.* In distinct contexts, people do ethnicity in different ways. In Toronto's Little Italy and Italian South Philadelphia the potential connections among people of Italian origins were actualized in distinct fashions; Italian ethnicity was a different kind of social practice.

While this perspective deliberately avoids questions of assimilation, I do not deny that such processes occurred. In postwar Canada and the United States, many people of Italian descent departed from ethnic communities, forging bonds at home, work, and play that were not predicated on ethnicity.[7] My research neither disputes this repeated sociological finding nor suggests that Philadelphia and Toronto proved exceptions to the rule. Many people of Italian descent in both postwar metropolitan areas stopped participating in the events that constituted Italian ethnicity. However, ethnicity continued to happen within enclaves at the heart of each city; urbanites made ethnicity an important facet of social life. The history of postwar cities such as Toronto and Philadelphia cannot be told without an understanding of how ethnicity operated in places like Little Italy and South Philadelphia. Irrespective of their impact on aggregate trends of assimilation—or indeed of ethnic retention—Italians, organized as such, contributed mightily to their postwar urban surroundings. Studies of the social meaning of urban change in the postwar era cannot exclude ongoing analysis of ethnicity.

These case studies should also contribute to the literature on the dynamic intersection of ethnicity and race. In his recent study, Thomas A. Guglielmo argues that Italians in Chicago saw racial categories shift in the first half of the twentieth century. Previously important racial distinctions—between northern and southern Italians, between all Italians and other European "races"—faded as race was increasingly reduced to a binary based on color.[8] The history of postwar South Philadelphia suggests that the black/white racial division carried powerful implications for the ways in which Italians conducted their daily lives. The social organization of Italian ethnicity into a neighborhood was part of the ongoing policing of a racial boundary. Comparison with Toronto demonstrates that Italian social life could take a dramatically different form in a context where the black/white binary held lesser importance. In Toronto, Italian ethnicity was a crucial source of social connection, but it was not organized to protect urban turf. Instead, Italian ethnicity took shape largely without concern for racial or ethnic boundary making. Race played a different role in the making of ethnicity in the two contexts.

But race was only one aspect of postwar urban environments that encouraged distinct practices of Italian ethnicity. Residents of Toronto's Little Italy and Italian South Philadelphia responded to opportunities and limits created by government policy and structure, the changing economies of the two cities, and the behavior of other city dwellers. In these regards, the story of postwar ethnicity differs greatly from that in the era of industrial growth. The uneven restructuring of postwar cities set new parameters for the development of urban ethnicity. Italians intimately felt changing urban structures in their everyday experiences. Postwar economic decline and the suburbanization of the city's white population were directly experienced by Italian South Philadelphians as their property values stagnated, parochial school classes shrank, and jobs shifted further from home. The very direct, personal threat posed by deindustrialization, along with prevailing schemes of racial categorization, infused urban experience with tensions that shaped ethnicity. In Toronto's Little Italy, rising property values, overflowing church processions, and ongoing economic opportunity gave concrete expression to the wider trajectory of the city. Economic prosperity and the structure of racial and ethnic relations reduced the need for territoriality. Economic and social structures outside the control of Italian residents played a crucial role in the making of ethnicity.

At the same time, I have maintained that Italians in each enclave made creative choices that constantly recast the mold of ethnic association. Territorial bonds in one sphere of activity encouraged territoriality in others; geographically elastic social ties in one arena fostered similar associations elsewhere. Residents of Toronto's Little Italy and Italian South Philadelphia fashioned their social lives from within the complex maze of opportunity and constraint that characterized postwar urban life in North America.[9]

Aspects of the story told here are national in scope. National migration histories shaped each local context. The mass arrival of Southern-born African Americans to cities in the Northern United States sets Philadelphia within a national narrative apart from Canadian cities.[10] As has been apparent throughout this study, immigration policies also played a crucial role in differentiating Toronto from Philadelphia.[11] Further, as Steven High argues, postwar economic restructuring was less devastating in southern Ontario than in the U.S. rustbelt because of differences in both government policy and labor movement tactics.[12]

Finally, the Canadian federal policy of multiculturalism—despite its ambiguous impact on ethnic experience more broadly—aided the development of large Italian institutions in Toronto in the 1970s and beyond. Government support of large community centers likely encouraged the Italian tendency to associate with coethnics outside of their neighborhoods.[13] However, the contrast between Canada and the United States should not be overdrawn; both countries played host to varied cities. Evidence from Toronto and Philadelphia demonstrates that locality matters in the making of social life; historians should exercise caution in generalizing from one local context to another.[14]

Study of Italians in Montreal warns against facile national generalizations. In his analysis of the city's Italian community in the 1960s, Jeremy Boissevain argues that the various Italian enclaves of Montreal constituted "isolated units which are almost as remote from each other as separate villages scattered across the countryside." Feasts and social activities were organized separately by the members of each neighborhood.[15] In an exploratory discussion of Italian social organization in Toronto and Montreal, Nicholas Harney observes "striking similarities" between the spatial arrangements of Italian community in the two cities, but also argues that key differences between the social and political dynamics of French and English Canada distinguish the two sites from one another.[16] Just as urban environments differ within any given national context, so too do the social practices of city residents.

Nonetheless, lessons gleaned from the study of Italians in Toronto and Philadelphia challenge previous characterizations of ethnic life. In both the United States and Canada, much of the historical research on Italian ethnicity—and ethnic communities more broadly—has focused on the prewar period. In the United States, the scarcity of social historical analysis of ethnicity in postwar cities has encouraged the notion that European ethnicity "fell latent" in mid-century America. Immigration restriction, the Great Depression, New Deal era unionism and politics, and the postwar "Consumer's Republic" all contributed, so the story goes, to dilution and eventual transformation of European ethnicity.[17] From this perspective, ethnicity ceased to be the kind of social bond that characterized neighborhoods as the English-speaking children and grandchildren of immigrants relocated to racially exclusive suburbs, where national origins might inform dietary preferences and church attendance but ethnicity no longer shaped lives. When ethnic identities regained political salience, they had been largely consolidated into a "white ethnic" constituency that could be mobilized in Philadelphia by figures such as Frank Rizzo and nationally as part of Richard Nixon's "silent majority."[18] The social history of South Philadelphia suggests a somewhat different relationship between older ethnic neighborhoods and the revival of ethnicity in the United States.

Hundreds of thousands (if not millions) of Americans continued to live in ethnic neighborhoods such as South Philadelphia in the decades after World War II.[19] Rather than breaking down and then reemerging in a new form, ethnicity continued to carry social organizational force throughout the twentieth century. This observation need not conflict with analysis of the emergence of "whiteness." In Philadelphia, historian Stefano Luconi is surely right to suggest that postwar racial politics in Philadelphia encouraged Italians to "reach out to other European groups" in the city.[20] However, the political consolidation of "whiteness" could coincide with the social practice of a more particularized ethnicity. Italians in Philadelphia came together with other white ethnics at the ballot box and in political protest while simultaneously keeping their real estate largely in Italian hands. "Whiteness" and ethnicity coexisted and cooperated. David Roediger notes this overlap as he comments that "the defense of the white neighbourhood from 'colored' outsiders became a defense of the 'ethnic neighborhood.'"[21] The evidence presented here suggests that this was precisely the case in South Philadelphia. Further social historical analysis of ethnic neighborhoods in postwar cities seems bound to confirm that ethnicity had greater importance in postwar cities than has been widely assumed.

In addition, the comparative dimension of this research should refine previous characterizations of the role of religion in the American "urban crisis." In their influential studies of postwar urban religion in the United States, Gerald Gamm and John T. McGreevy use longstanding theological and institutional characteristics of Catholicism to explain the responses of Catholics to postwar urban change. In their view, Catholics were more likely than Protestants and Jews to remain in their old neighborhoods because of the social and institutional organization of Catholicism.[22] As Gamm puts it: "Ancient rules binding churches and synagogues have shaped the twentieth-century urban battle of race and housing. . . . institutions were defined, bound, and ultimately constrained by rules that dictated their own inexorable logic."[23] However, evidence of Catholic territorialism in postwar American cities provides only limited support for the suggestion that "ancient rules" had the capacity to "dictate" Catholic behavior.

To be sure, many Catholics in postwar American cities, including Italians in South Philadelphia, defended urban turf. In early postwar Chicago, mobs composed of "an 'ethnic' amalgam of working-class

Catholics" mobilized to prevent black residents from purchasing houses in their neighborhoods.[24] Similarly, Roman Catholic areas in Detroit were especially likely to mount violent resistance to prospective African American neighbors, while Irish and Italian Catholics in Boston resisted bussing policies designed to promote school integration.[25] Territorial Catholics, willing to resort to violence to protect their neighborhoods, are frequent players in postwar urban history.[26] This study expands the literature on Catholic territorialism in the postwar United States by demonstrating the importance of locality to less explicitly politicized activities, such as real estate transactions, marriage choices, and church participation.

However, the comparative dimension of this research suggests that territorialism was not an inherent characteristic of Catholic urban life in North America. Rather, as Robert Orsi suggests, urban Catholicism can only be understood "situationally . . . not within the terms of a religious tradition . . . understood as existing apart from history."[27] Accordingly, I have argued that Catholic territorialism in Philadelphia did not flow inexorably from the ancient rules of Catholicism but rather from decisions of Catholics who operated within the constraints of urban conflict. In Toronto, where political, economic, and demographic patterns yielded a different context, Catholic life assumed a contrasting spatial form.[28] The geographical content of Italian ethnic life and the "rules" of Catholicism varied in accordance with their local urban contexts.

At the same time, this study aims to push social historical analysis of ethnicity in Canada into a new stage. Significant scholarly attention has demonstrated the importance of immigrant enclaves to urban life in late nineteenth- and early twentieth-century Canada. Whether sojourning laborers or permanent settlers, immigrants in the industrial heyday concentrated in urban neighborhoods. Often, they were pushed into residential clusters by the prejudices of other Canadians, who then decried the supposed environmental and moral ills of immigrant enclaves. But for immigrants, the enclaves offered considerable advantages. Immigrants gathered close to sources of employment, and their geographic concentration encouraged social and economic connections that eased the transition to Canada.[29] But in the postwar era, the relationship between urban social organization and ethnicity in Canada grows murky.

For Italians in Toronto, evidence suggests dramatic changes from the prewar to postwar eras. Little Italy began as a neighborhood. The first Italians in the area came from Conzena in Italy's southern region of Calabria. They concentrated together because of the close social and economic connections among emigrants from the same hometowns.[30] In the first quarter of the twentieth century, the Calabrese in Little Italy were joined by emigrants from other Italian regions. During this period, Italians in the area surrounding College Street developed close social ties with fellow ethnics in the neighborhood. Bonds among people from different *paese* were fostered by shared streets, workplaces, institutions, and houses.[31] As neighborhood assumed importance, tension arose among the various Italian enclaves in the city. Few residents relocated from Little Italy to another area of Italian settlement in the city, and residents of each Italian neighborhood expressed suspicion and hostility toward the others.[32]

The leading scholar of Italian life in the postwar era, Franca Iacovetta, describes the experiences of postwar immigrants in the city but does not focus on the changing social geography of ethnic life. Iacovetta rightly reports that postwar Italian immigrants gravitated to areas settled by previous generations of Italians in city. Her suggestion that Italian newcomers to Toronto in the 1950s limited their movement to "a fairly narrow geographical area" follows from the prewar literature on this topic, which suggests that the various Italian settlements in the city constituted socially separate neighborhoods.[33] Yet, a transformation of Italian life in Toronto began in the 1950 and 1960s. By the 1970s, Italian ethnicity in Toronto scarcely retained the insularity of the industrial heyday. The shape of Italian ethnic life in Toronto—itself a response to urban structures—ultimately resulted in the dissipation of the old residential neighborhood.

Most likely, further research will reveal other groups that mirrored the patterns displayed by Italians in Toronto. Historic "Chinatowns" have remained centers of Chinese social and economic activity in Canadian cities, despite the dispersal of many Chinese ethnics across urban and suburban landscapes.[34] Greek and Portuguese immigrants have similarly preserved space at the center of cities while dispersing into neighborhoods across metropolitan areas.[35] A study of Japanese Canadians in postwar Vancouver has suggested that Taiko drumming festivals allowed members of that community to gather in the sections of the city from which they were forcibly removed during World War II.[36] For these ethnic groups, among others, commercial, cultural, and social activities at the urban core may unify dispersed communities. Events that draw geographic outsiders together in a central city area may play a crucial role in postwar Canadian urban experience.

As these strands suggest, the history of Canadian ethnic groups after World War II—the most ethnically diverse chapter of Canadian history—has yet to be adequately connected with postwar urban history. And yet, as Daniel Hiebert notes in his exploratory social geography of Vancouver, "Canadian multiculturalism is being defined in places like metropolitan Vancouver, where groups from around the world interact in the local economy and in the neighbourhoods they create."[37] With more than 60 percent of Canadian immigrants residing Toronto, Vancouver, and Montreal, the story of Canadian diversity has truly become a story of cities.[38] At the same time, the history of urban Canada cannot be told without careful attention to the activities of immigrants.

The ways that ethnic urbanites practiced ethnicity depended on who they were and where they were. As urban environments differed from place to place and changed over time they required dynamic responses from city residents. Comparison of Italian ethnicity in postwar Toronto and Philadelphia confirms the view that ethnicity is a practice rather than an attribute, and highlights the role of urban historical variation in shaping local ethnic practices.

Italian ethnicity happened differently in postwar Toronto and Philadelphia. In each city, ethnic social bonds took their own, surprising forms. In the choreography of their daily lives, residents of Toronto's Little Italy and Italian South Philadelphia gave expression to the intimate reverberations of urban change and to the creative possibilities of social experience.

Notes

INTRODUCTION

1. Annunciation parishioners, group interview by author, Philadelphia, Pennsylvania, May 10, 2004.

2. Audrey Geniole, interview by A. McPeek, Toronto, ON, September 20, 1977, Multicultural History Society of Ontario (MHSO).

3. Audrey Geniole, interview by A. McPeek, Toronto, ON, September 20, 1977, MHSO.

4. Three notes on diction: (1) Throughout this book, I use the terms "Italian," "Italian ethnic," and "Italian origins" synonymously to designate both people of Italian birth and people whose ancestry traces to Italy. I specify whenever I mean a subset of this larger group. This choice is made for semantic purposes. My conception of ethnic bonds is elaborated in the conclusion. (2) I use the phrase "Italian South Philadelphia" to describe the portion of South Philadelphia that includes the two parishes (Annunciation and St. Thomas) that are the focus of my Philadelphia research, as well as the larger surrounding area populated predominantly by people of Italian extraction. This area is described in chapter 1 and pictured in map 4. (3) I use "Little Italy" to describe the Italian residential enclave centered at College and Grace in Toronto. This choice reflects common usage in Toronto. It is worth noting, however, that were always other, and eventually larger, concentrations of Italians in the metropolitan area, as described in chapter 1.

5. This history is elaborated in chapter 1; see also Robert F. Harney and J. Vincenza Scarpaci, eds., *Little Italies in North America* (Toronto: Multicultural History Society of Ontario, 1981).

6. There is a massive and still growing literature on the economic and racial dynamics of post–World War II American cities. Some exemplary examples include Matthew J. Countryman, *Up South: Civil Rights and Black Power in Philadelphia* (Philadelphia: University of Pennsylvania, 2006); Matthew D. Lassiter, *The Silent Majority: Suburban Politics in the Sunbelt South* (Princeton: Princeton University Press, 2006); Kevin Kruse, *White Flight: Atlanta and the Making of Modern Conservatism* (Princeton: Princeton University Press, 2005); Amanda I. Seligman, *Block by Block: Neighborhoods and Public Policy on Chicago's West Side* (Chicago: University of Chicago Press, 2005); Robert O. Self, *American Babylon: Race and the Struggle for Postwar Oakland* (Princeton: Princeton University Press, 2003); Thomas J. Sugrue, *The Origins of the Urban Crisis: Race and Inequality in Postwar Detroit* (Princeton: Princeton University Press, 1996); Douglas S. Massey and Nancy A. Denton, *American Apartheid: Segregation and the Making of the Underclass* (Cambridge, MA: Harvard University Press, 1993); Thomas J. Sugrue, "The Structures of Urban Poverty: The Reorganization of Space and Work in Three Periods of American History," in *The 'Underclass' Debate: Views from History,* ed. Michael B. Katz, 85–117 (Princeton: Princeton University Press, 1993); Adams et al., *Philadelphia;* John T. Cumbler, *A Social History of Economic Decline: Business, Politics, and Work in Trenton* (New Brunswick, NJ: Rutgers University Press, 1989); Kenneth T. Jackson, *Crabgrass Frontier: The Suburbanization of the United States* (New York: Oxford University Press, 1985); Arnold R. Hirsch, *Making the Second Ghetto: Race and Housing in Chicago, 1940–1960* (1983; Chicago: University of Chicago Press, 1998); and David M. Gordon, Richard Edwards, and Michael Reich, *Segmented Work, Divided Workers: The Historical Transformation of Labor in the United States* (New York: Cambridge University Press, 1982).

7. For typologies of postwar cities, see Saskia Sassen, "New Inequalities among Cities," and "The New Urban Economy: The Intersection of Global Processes and Place," in *Cities in a World Economy,* 2nd ed. (Thousand Oaks, CA: Pine Forge Press, 2000), chaps. 3–4; H. V. Savitch and Paul Kantor, *Cities in the International Marketplace: The Political Economy of Urban Development in North America and Western Europe* (Princeton: Princeton University Press, 2002).

8. Massey and Denton, *American Apartheid.* For case studies of social and economic division in sites of dramatic decline and significant prosperity, see Sugrue, *Origins of the Urban Crisis;* and Mike Davis, *City of Quartz: Excavating the Future in Los Angeles* (New York: Vintage, 1992).

9. These figures include individuals who listed "Italian" and "other" origins. Italians were not alone in preserving ethnic city space. More than 125,000 people with Irish ancestry, 134,000 with Polish ancestry, and (startlingly) 4,000,000 with German ancestry continued to live as a majority in their own census tracts in 1980. I am not suggesting a parallel to African American spatial isolation in the midcentury United States. As a point of comparison, in 1980 almost 15,000,000 African Americans lived as a majority in their census tracts. Data accessed at http://www.nhgis.org/ (June 16, 2008). For sociological studies that have emphasized the weakening of European

national origins groups in mid-century America, see Mary C. Waters, *Ethnic Options: Choosing Identities in America* (Berkeley: University of California Press, 1990); Stanley Lieberson and Mary C. Waters, *From Many Strands: Ethnic and Racial Groups in Contemporary America* (New York: Russell Sage Foundation, 1988); Richard Alba, *Italian Americans: Into the Twilight of Ethnicity* (Englewood Cliffs, NJ: Prentice-Hall, 1985); and Herbert Gans "Symbolic Ethnicity: The Future of Ethnic Groups and Cultures in America," *Ethnic and Racial Studies* 2 (1979): 1–20.

10. This study thus seeks to build upon, but also to refine, the observations of a variety of scholars who have suggested the displacement of European national origins groups by overarching white racial affiliations. See, for example, Matthew Frye Jacobson, *Whiteness of a Different Color: European Immigrants and the Alchemy of Race* (Cambridge, MA: Harvard University Press, 1999), 91–135; David R. Roediger, *Working Toward Whiteness: How America's Immigrants Became White; The Strange Journey from Ellis Island to the Suburbs* (New York: Basic Books, 2005), 133–234; Lizabeth Cohen, *Making a New Deal: Industrial Workers in Chicago, 1919–1939* (New York: Cambridge University Press, 1990), idem, *A Consumers' Republic: The Politics of Mass Consumption in Postwar America* (New York: Alfred Knopf, 2003); Gary Gerstle, *Working-Class Americanism: The Politics of Labor in a Textile City, 1914–1960* (Cambridge: Cambridge University Press, 1989); and Steve Fraser and Gary Gerstle, *The Rise and Fall of the New Deal Order, 1930–1980* (Princeton: Princeton University Press, 1989). To a significant extent, my efforts here run in concert with those of Joshua M. Zeitz, *White Ethnic New York: Jews, Catholics and the Shaping of Postwar Politics* (Chapel Hill: University of North Carolina Press, 2007); however, to a greater extent than Zeitz, I see race and ethnicity as complementary frameworks for the analysis of postwar urban history.

11. Gerald H. Gamm, *Urban Exodus: Why the Jews Left Boston and the Catholics Stayed* (Cambridge, MA: Harvard University Press, 1999); John T. McGreevy, *Parish Boundaries: The Catholic Encounter with Race in the Twentieth-Century Urban North* (Chicago: University of Chicago Press, 1996), 19–28; Eileen M. McMahon, *What Parish Are You From?: A Chicago Irish Community and Race Relations* (Lexington: University Press of Kentucky, 1995); Sugrue, *Origins of the Urban Crisis,* 235–258; Ronald P. Formisano, *Boston Against Bussing: Race, Class, and Ethnicity in the 1960s and 1970s* (Chapel Hill: University of North Carolina Press, 1991); Hirsch, "Friends, Neighbors, and Rioters," in *Making the Second Ghetto,* 68–99; Jonathan Rieder, *Canarsie: The Jews and Italians of Brooklyn against Liberalism* (Cambridge, MA: Harvard University Press, 1987), 32–43; Robert A. Orsi, *The Madonna of 115th Street: Faith and Community in Italian Harlem, 1880–1950,* 2nd ed. (New Haven: Yale University Press, 2002), 33–36

12. Orsi, *Madonna of 115th Street,* xix (emphasis in the original).

13. James T. Lemon, *Toronto, since 1918: An Illustrated History* (Toronto: James Lorimer, 1985); Statistics Canada, Historical Statistics of Canada, table A385–416: "Immigration to Canada by Country of Last Permanent Residence, 1956–1976," http://www.statcan.gc.ca/pub/11-516-x/sectiona/

4147436-eng.htm (accessed February 10, 2009); and Harold Troper, "Becoming an Immigrant City: A History of Immigration to Toronto since World War II," in *The World in a City,* ed. Paul Anisef and C. Michael Lanphier, 19–62 (Toronto: University of Toronto Press, 2003).

14. Graham Todd, "'Going Global' in the Semi-periphery: World Cities as Political Projects; The Case of Toronto," in *World Cities in a World-System,* ed. Paul L. Knox and Peter J. Taylor (New York: Cambridge University Press, 1995); Sassen, *Cities in a World Economy;* and Donald P. Kerr, "The Economic Structure of Toronto," in *Toronto,* ed. Jacob Spelt and Donald P. Kerr (Toronto: Collier-MacMillan Canada, 1973).

15. Lemon, *Toronto, since 1918,* 135; Timothy J. Colton, *Big Daddy: Frederick G. Gardiner and the Building of Metropolitan Toronto* (Toronto: University of Toronto Press, 1980). On the importance of metropolitan government, and definitions thereof, to urban development and health, see Savitch and Kantor, *Cities in the International Marketplace;* and Gerald E. Frug, *City Making: Building Communities without Building Walls* (Princeton: Princeton University Press, 1999).

16. Statistics Canada, Historical Statistics of Canada, table A385–416: "Immigration to Canada by Country of Last Permanent Residence, 1956–1976," http://www.statcan.gc.ca/pub/11-516-x/sectiona/4147436-eng.htm [Accessed February 10, 2009]; Canadian Census Data, census tract level from 1971, accessed June 19, 2008 from the University of Toronto Computing in the Humanities and Social Sciences (http://www.chass.utoronto.ca/cgi-bin/chassnew/display.pl?page=index).

17. On the geographic paths of social life and the "choreography" of social experience, see Torsten Hagerstrand, "What about People in Regional Science?" *Regional Science Association Papers* 24 (1970): 7–21; Allan Pred, "The Choreography of Existence: Comments on Hagerstrand's Time-Geography and Its Usefulness," *Economic Geography* 53, no. 2 (April 1977): 207–221; Anthony Giddens, "Time, Space and Regionalization," in *The Constitution of Society: Outline of the Theory of Structuration* (Berkeley: University of California Press, 1984), chap. 3; and Derek Gregory, *Geographic Imaginations* (Cambridge, MA: Blackwell, 1994), chap. 2. On the notion of geographically elastic urban communities, see Wilbur Zelinsky and Barrett A. Lee, "Heterolocalism: An Alternative Model of the Sociospatial Behavior of Immigrant Ethnic Communities," *International Journal of Population Geography* 4, no. 4 (December 1998): 281–298; Wilbur Zelinsky, *The Enigma of Ethnicity: Another American Dilemma* (Iowa City: University of Iowa City Press, 2001); Kenneth A. Scherzer, *The Unbounded Community: Neighborhood Life and Social Structure in New York City, 1830–1875* (Durham: Duke University Press, 1992); M. M. Webber, "Order in Diversity: Community without Propinquity," in *Cities and Space: The Future Use of Urban Land,* ed. L. Wingo Jr., 23–54 (Baltimore: Johns Hopkins University Press, 1963).

18. Kathleen Neils Conzen et al., "The Invention of Ethnicity: A Perspective from the USA," *Journal of American Ethnic History* 12, no. 1 (Fall 1992): 3–41; William L. Yancey, Eugene P. Ericksen, and Richard N. Juliani, "Emer-

gent Ethnicity: A Review and Reformulation," *American Sociological Review* 41, no. 3 (1976): 391–403.

19. William Sites, "Global City, American City," *Journal of Urban History* 29, no. 3 (March 2003): 333–346; Sassen, "New Inequalities among Cities," chap. 3; John Friedmann, "Where We Stand: A Decade of World City Research," in Knox and Taylor, *World Cities in a World-System,* 21–47.

CHAPTER ONE

1. William Sites, "Global City, American City," *Journal of Urban History* 29, no. 3 (March 2003): 333–346; Saskia Sassen, "New Inequalities among Cities," and "The New Urban Economy: The Intersection of Global Processes and Place," in *Cities in a World Economy,* 2nd ed. (Thousand Oaks, CA: Pine Forge Press, 2000), chaps. 3–4; John Friedmann, "Where We Stand: A Decade of World City Research," in *World Cities in a World-System,* ed. Paul L. Knox and Peter J. Taylor, 21–47 (New York: Cambridge University Press, 1995); Phillip Blumberg, *The Megacorporation in American Society* (Englewood Cliffs, NJ: Prentice-Hall, 1975); Michael Best, *The New Competition: Institutions of Industrial Restructuring* (Cambridge, MA: Harvard University Press, 1990); Carl Abbott, *The New Urban America: Growth and Politics in Sunbelt Cities* (Chapel Hill: University of North Carolina Press, 1981); John H. Mollenkopf, *The Contested City* (Princeton: Princeton University Press, 1983); Ann Markusen et al., *The Rise of the Gunbelt: The Military Remapping of Industrial America* (New York: Oxford University Press, 1991); Paul Kantor and Steven David, *The Dependent City: The Changing Political Economy of Urban America* (Boston: Scott, Foresman, and Co., 1988), 165–172; and H. V. Savitch and Paul Kantor, *Cities in the International Marketplace: The Political Economy of Urban Development in North America and Western Europe* (Princeton: Princeton University Press, 2002).

2. John Friedmann, "The World City Hypothesis," *Development and Change* 17, no. 1, (1986): 69–84; Nigel Thrift, "The Fixers: The Urban Geography of International Commercial Capital," in *Global Restructuring and Territorial Development,* ed. J. Henderson and Manuel Castells (London: Sage, 1987); Graham Todd, "'Going Global' in the Semi-periphery: World Cities as Political Projects; The Case of Toronto," in Knox and Taylor, *World Cities in a World-System;* Sassen, *Cities in a World Economy,* 92–95. A recent examination of Philadelphia's economic and demographic history suggests the extent to which, compared with other cities, it continued to struggle even after the improvements of the 1980s. See Daniel Amsterdam, "Immigration to the City of Philadelphia: An Economic and Historical Overview" (Philadelphia Migration Project Working Paper, 2007), accessed at http://www.history.upenn.edu/philamigrationproject/papers.html.

3. Dominion Bureau of Statistics, *1961 Census of Canada,* Series CT, "Population and Housing Characteristics by Census Tracts," Toronto, table 2, 26 (Ottawa: Ministry of Trade and Commerce, 1964); Statistics Canada, *1971 Census of Canada,* Census Tract Bulletin, Series B, "Population and Hous-

ing Characteristics by Census Tract," Toronto, table 2, 50 (Ottawa: Ministry of Trade and Commerce, 1974); Statistics Canada, *1981 Census of Canada,* Profile Series B, vol. 3, Census Metropolitan Areas with Components, Selected Social and Economic Characteristics, table 1 (Ottawa: Ministry of Trade and Commerce, 1983); Statistics Canada, *1991 Census of Canada,* Profile of Census Divisions and Subdivisions in Ontario, Part B, table 1 (Ottawa: Ministry of Industry, 1994). On the history of working-class suburbs in Toronto, see Richard Harris, *Unplanned Suburbs: Toronto's American Tragedy, 1900–1950* (Baltimore: Johns Hopkins University Press, 1996).

4. 1980 U.S. Census Extract, County Level Data (Pennsylvania: Philadelphia, Bucks, Montgomery, Chester, and Delaware Counties; New Jersey: Burlington, Camden, and Gloucester Counties), NT39: "Median Value of Specified Owner-Occupied Noncondominium Housing Units," National Historical Geographic Information Systems, Minnesota Population Center, University of Minnesota, www.nhgis.org (accessed October 2008). On the economic disparity between Philadelphia and the wider region in the postwar period, see Carolyn Adams et al., *Philadelphia: Neighborhoods, Division, and Conflict in a Postindustrial City* (Philadelphia: Temple University Press, 1991).

5. Ninette Kelly and Michael Trebilcock, *The Making of the Mosaic: A History of Canadian Immigration Policy* (Toronto: University of Toronto Press, 1998), chaps. 8–9; Valerie Knowles, *Strangers at Our Gates: Canadian Immigration and Immigration Policy, 1540–2006* (Toronto: Dundurn Press, 2007), chaps. 8–9; Roger Daniels, "The Cold War and Immigration," and "Lyndon Johnson and the End of the Quota System," in *Guarding the Golden Door: American Immigration Policy and Immigration since 1882* (New York: Hill and Wang, 2004), chaps. 6–7; George J. Borjas, "Immigration Policy, National Origin, and Immigration Skills: A Comparison of Canada and the United States," in *Small Differences That Matter: Labor Markets and Income Maintenance in Canada and the United States,* ed. David Card and Richard B. Freeman, 21–43 (Chicago: University of Chicago Press, 1993).

6. Daniels, *Guarding the Golden Door,* 123; Statistics Canada, Historical Statistics of Canada, table A350: "Immigrant Arrivals to Canada 1852–1977," http://www.statcan.ca/english/freepub/11-516-XIE/sectiona/sectiona.htm#Immigration (accessed July 19, 2004).

7. James T. Lemon, *Toronto, since 1918: An Illustrated History* (Toronto: James Lorimer, 1985), 196; Statistics Canada, Historical Statistics of Canada, table A385–416: "Immigration to Canada by Country of Last Permanent Residence, 1956–1976," http://www.statcan.ca/english/freepub/11-516-XIE/sectiona/A385_416.csv (accessed March 11, 2005); Harold Troper, "Becoming an Immigrant City: A History of Immigration into Toronto since World War II," in *The World in a City,* ed. Paul Anisef and C. Michael Lanphier, 19–62 (Toronto: University of Toronto Press, 2003).

8. University of Pennsylvania Library, *1970 Census Tract Data of Philadelphia,* extracted from the Contextual Data Archive of Sociometrics Corporation, http://data.library.upenn.edu/phila.html (accessed on March 11, 2005).

9. Commission on Human Relations, "Philadelphia's Negro Population, Facts on Housing" (Philadelphia: Commission on Human Relations, 1953), 1;

National Historical Geographic Information Systems, Minnesota Population Center, University of Minnesota, www.nhgis.org (accessed October 2008).

10. U.S. Census Bureau, "Population of the 100 Largest Cities and Other Urban Places"; Adams et al., *Philadelphia,* 9 and 18. In Burlington County, the black population in 1980 comprised 15 percent of the total; in Camden County the figure was 19 percent. On the Great Migration, see Nicholas Lemann, *The Promised Land: The Great Black Migration and How It Changed America* (New York: Vintage, 1991); James R. Grossman, *Land of Hope: Chicago, Black Southerners, and the Great Migration* (Chicago: University of Chicago Press, 1989); and Reynolds Farley and Walter R. Allen, *The Color Line and the Quality of Life in America* (New York: Russell Sage Foundation, 1987), 103–159.

11. Kelly and Trebilcock, *Making of the Mosaic,* chaps. 9 and 10; Lemon, "Multicultural and Financial Metropolis, 1955–1984," in *Toronto, since 1918,* chap. 5; Troper, "Becoming an Immigrant City."

12. Daniels, *Guarding the Golden Door,* 5; U.S. Census Bureau, *1990 Summary Tape File* (STF-3), Philadelphia Metropolitan Statistical Area, http://factfinder.census.gov/servlet/DatasetMainPageServlet?_ds_name=DEC_1990_STF1_&_program=DEC&_lang=en (accessed March 11, 2005). This is not to say that immigration played no role in Philadelphia's postwar history. For a treatment of new immigrant groups in Philadelphia and their capacity to complicate racial dynamics, see Judith Goode and Jo Anne Schneider, *Reshaping Ethnic and Racial Relations in Philadelphia: Immigrants in a Divided City* (Philadelphia: Temple University Press, 1994).

13. Kenneth John Rea, *The Prosperous Years: The Economic History of Ontario, 1939–1975* (Toronto: University of Toronto Press, 1985), 42–43, 193–222.

14. Author's calculations, based on Rea, *Prosperous Years,* 196; Lemon, *Toronto, since 1918,* 197.

15. Donald P. Kerr, "The Economic Structure of Toronto," in *Toronto,* ed. Jacob Spelt and Donald P. Kerr (Toronto: Collier-MacMillan Canada, 1973), 55.

16. Lemon, *Toronto, since 1918,* 197.

17. Donald P. Kerr, "Economic Structure of Toronto," 70.

18. Todd, "'Going Global,'" 197–198.

19. Gunter Gad, "Toronto's Central Office Complex: Growth, Structure and Linkages" (Ph.D. thesis, University of Toronto, 1975); Todd, "'Going Global,'" 202.

20. Todd, "'Going Global,'" 200.

21. Anita A. Summers and Thomas F. Luce, *Economic Development within the Philadelphia Metropolitan Area* (Philadelphia: University of Pennsylvania Press, 1987), 210, 221; Philip Scranton, "Large Firms and Industrial Restructuring: The Philadelphia Region, 1900–1980," *Pennsylvania Magazine of History and Biography* 116, no. 4 (1992): 420.

22. Scranton, "Large Firms and Industrial Restructuring"; Adams et al., "Economic Erosion and the Growth of Inequality," in *Philadelphia,* chap. 2.

23. William J. Still and Janice Fanning Madden, *Post-Industrial Philadel-*

phia: Structural Changes in the Metropolitan Economy (Philadelphia: University of Pennsylvania Press, 1990), 44–66.

24. Summers and Luce, *Economic Development,* 210, 221.

25. Theodore Hershberg, *At the Crossroads: The Consequences of Economic Stability or Decline in Philadelphia for the City, Region and Commonwealth* (Philadelphia: Center for Greater Philadelphia, 1991); Scranton, "Large Firms and Industrial Restructuring," 419–420.

26. Still and Madden, *Post-Industrial Philadelphia,* 12; Summers and Luce, *Economic Development,* 210.

27. John D. Kasarda, "Urban Change and Minority Opportunities," in *The New Urban Reality,* ed. Paul E. Peterson, 45 (Washington, DC: Brookings Institution, 1985).

28. H. Carl Goldenberg, *Report of the Royal Commission on Metropolitan Toronto* (Toronto: Royal Commission on Metropolitan Toronto, 1965), 26–37; Ronald C. Smith et al., "The Organization of Local Government in Metropolitan Toronto," report prepared for the Royal Commission on Metropolitan Toronto (1975), 19–31; John P. Robarts, *Report of the Royal Commission on Metropolitan Toronto,* vol. 1 (Toronto: Royal Commission on Metropolitan Toronto, 1977), 13–17; Lemon, *Toronto, since 1918,* 135. On the influential first chairman of Metro, see Timothy J. Colton, *Big Daddy: Frederick G. Gardiner and the Building of Metropolitan Toronto* (Toronto: University of Toronto Press, 1980).

29. Robarts, *Report of the Royal Commission on Metropolitan Toronto,* 19.

30. George A. Nader, *Cities of Canada: Profiles of Fifteen Metropolitan Centres* (Toronto: MacMillan, 1976), 191. On the importance of metropolitan government, and definitions thereof, to urban development and health, see Savitch and Kantor, *Cities in the International Marketplace;* Gerald E. Frug, *City Making: Building Communities without Building Walls* (Princeton: Princeton University Press, 1999).

31. Lemon, *Toronto, since 1918.*

32. Jim Simmons "Commercial Structure and Change in Toronto," Research Paper 182 (Toronto: University of Toronto Department of Geography and Centre for Urban and Community Studies, 1991); J. W. Simmons and A. Baker, "Household Relocation Patterns," and B. Wellman et al., "Community Ties and Support Systems: From Intimacy to Support," both in *The Form of Cities in Central Canada: Selected Papers,* ed. L. S. Bourne, R. D. MacKinnon, and J. W. Simmons, 199–217, 156–162 (Toronto: University of Toronto Press, 1973).

33. Bureau of Municipal Research, "In Response to the Robarts Report," *Topic* 2 (October 1977), 7 (emphasis in the original).

34. Adams et al., *Philadelphia.* On this topic more broadly, see Richard Dilworth, *The Urban Origins of Suburban Autonomy* (Cambridge, MA: Harvard University Press, 2005); Myron Orfield, *American Metropolitics: The New Suburban Reality* (Washington, DC: Brookings Institution Press, 2001); Frug, *City Making;* Elinor Ostrom, "The Social Stratification-Government Inequality Thesis Explored," *Urban Affairs Quarterly* 19, no. 1 (1983): 91–112. For an explicit comparison of city structure in the United States and Canada,

see Michael A. Goldberg and John Mercer, *The Myth of the North American City: Continentalism Challenged* (Vancouver: University of British Columbia Press, 1986).

35. For an excellent summary of the political economy of postwar Philadelphia with a somewhat different focus, see Matthew J. Countryman, *Up South: Civil Rights and Black Power in Philadelphia* (Philadelphia: University of Pennsylvania Press, 2006), 48–79. For the growing literature on this topic more broadly, see Matthew D. Lassiter, *The Silent Majority: Suburban Politics in the Sunbelt South* (Princeton: Princeton University Press, 2006); Kevin Kruse, *White Flight: Atlanta and the Making of Modern Conservatism* (Princeton: Princeton University Press, 2005); Amanda I. Seligman, *Block by Block: Neighborhoods and Public Policy on Chicago's West Side* (Chicago: University of Chicago Press, 2005); Robert O. Self, *American Babylon: Race and the Struggle for Postwar Oakland* (Princeton: Princeton University Press, 2003); Thomas J. Sugrue, *The Origins of the Urban Crisis: Race and Inequality in Postwar Detroit* (Princeton: Princeton University Press, 1996); Douglas S. Massey and Nancy A. Denton, *American Apartheid: Segregation and the Making of the Underclass* (Cambridge, MA: Harvard University Press, 1993); Thomas J. Sugrue, "The Structures of Urban Poverty: The Reorganization of Space and Work in Three Periods of American History," in *The "Underclass" Debate: Views from History,* ed. Michael B. Katz, 85–117 (Princeton: Princeton University Press, 1993); Adams et al., *Philadelphia;* John T. Cumbler, *A Social History of Economic Decline: Business, Politics, and Work in Trenton* (New Brunswick, NJ: Rutgers University Press, 1989); Kenneth T. Jackson, *Crabgrass Frontier: The Suburbanization of the United States* (New York: Oxford University Press, 1985); Arnold R. Hirsch, *Making the Second Ghetto: Race and Housing in Chicago* (1983: Chicago: University of Chicago Press, 1998); David M. Gordon, Richard Edwards, and Michael Reich, *Segmented Work, Divided Workers: The Historical Transformation of Labor in the United States* (New York: Cambridge University Press, 1982).

36. Commission on Human Relations, *Racial Discrimination in Housing: Findings and Recommendations* (Philadelphia: Commission on Human Relations, 1966), 8; "Investigation into Alleged Unethical Soliciting Practices by Real Estate Agents and Brokers," Public Hearing February 17, 1960, box A-2843, group 148, Commission on Human Relations Records, Philadelphia City Archives (PCA); "Public Investigatory Hearings into Discrimination in Housing," Public Hearings November 1965, box A-2844, group 148, Commission on Human Relations Records, PCA. For the creation of the commission, see Pennsylvania General Assembly, "Philadelphia Home Rule Charter," April 17, 1951, Harrisburg, reprint, 1991, 62–63. The Pennsylvania state legislature elaborated the role of the commission in two subsequent acts, the "Pennsylvania Fair Employment Practice Act" of 1955 and its amended version in 1961. See Pennsylvania General Assembly, *Laws of the General Assembly of the Commonwealth of Pennsylvania, 1961* V 1 (Harrisburg, PA: 1961), 47–50.

37. "Clinical Inquiry into Incidents of Intimidating and Mob Pressure Forcing Minority Families to Vacate Their Homes," December 19, 1960, box

A-2843, group 148, Commission on Human Relations Records, PCA, 15–16, and Exhibit C-6, "Statement of Mrs. Maximina Figueroa to the Commission on Human Relations."

38. "Clinical Inquiry into Incidents of Intimidating and Mob Pressure Forcing Minority Families to Vacate Their Homes," December 19, 1960, box A-2843, group 148, Commission on Human Relations Records, PCA, 52, 10, and 76.

39. Commission on Human Relations, *The Housing of Negro Philadelphians* (Philadelphia: Commission on Human Relations, 1953).

40. Massey and Denton, *American Apartheid,* 64.

41. Afua Cooper, *The Hanging of Angelique: The Untold History of Canadian Slavery and the Burning of Old Montreal* (Toronto: Harper Collins, 2006); Kristin McLaren, "'We Had No Desire to Be Set Apart': Forced Segregation of Black Students in Canada West Public Schools and Myths of British Egalitarianism," *Social History/Histoire Sociale* 37, no. 73 (2004): 27–50; Barrington Walker, "The Gavel and the Veil of Race: 'Blackness' in Ontario's Criminal Courts, 1858–1958" (Ph.D. thesis, University of Toronto, 2003); Constance Backhouse, *Colour-Coded: A Legal History of Racism in Canada, 1900–1950* (Toronto: University of Toronto Press, 1999); James W. St. G. Walker, *"Race," Rights and the Law in the Supreme Court of Canada: Historical Case Studies* (Waterloo, ON: Wilfred University Press, 1997); Robin W. Winks, *The Blacks in Canada: A History,* 2nd ed. (Montreal: McGill-Queen's University Press, 1997); James W. St. G. Walker, *The Black Loyalists: The Search for a Promised Land in Nova Scotia and Sierra Leone, 1783–1870,* 2nd ed. (Toronto: University of Toronto Press, 1992), idem, *Discrimination in Canada: The Black Experience* (Ottawa: Canadian Historical Association 1985).

42. University of Toronto Data Library Service, *Census of Canada, 1971: Public Use Microdata on Individuals—Toronto and Montreal Sample,* Census of Canada Microdata, http://r1.chass.utoronto.ca.myaccess.library.utoronto.ca/cgi-bin/sda/hsda?harcsda+cc71ic (accessed on March 11, 2005). Figure based on the "negro" category in the "ethnic or cultural group" variable. On racism in postwar Toronto, see Frances Henry, *The Caribbean Diaspora in Toronto: Learning to Live with Racism* (Toronto: University of Toronto Press, 1994); Keith D. Lowe, *Race Relations in Metropolitan Toronto, 1982: A Situation Report* (Ottawa: Multiculturalism Directorate, 1982); Leon Muszynski and Jeffrey Reitz, *Racial and Ethnic Discrimination in Employment,* Working Papers for Full Employment 5 (Toronto: Social Planning Council of Metropolitan Toronto, 1982); Frances Henry, *The Dynamics of Racism in Toronto: Research Report* (Ottawa: Secretary of State of Canada, 1978); and Wilson A. Head, *The Black Presence in the Canadian Mosaic: A Study of Perception and the Practice of Discrimination Against Blacks in Metropolitan Toronto* (Toronto: Ontario Human Rights Commission, 1975).

43. Author's calculations, using entire Census Metropolitan Area. See http://citystats.uvic.ca/.

44. John Myles and Feng Hou, "Changing Colours: Spatial Assimilation and New Racial Minority Immigrants," *Canadian Journal of Sociology* 29, no.

1 (2004): 29–58; Robert A. Murdie and Carlos Teixeira, "Towards a Comfortable Neighborhood and Appropriate Housing: Immigrant Experiences in Toronto," in Anisef and Lanphier, *World in a City*, 132–191; D. Ley, *Is There an Immigrant 'Underclass' in Canadian Cities?* RIIM, Working Paper Series no. 97–108 (Vancouver: Vancouver Centre of Excellence 1997); Robert A. Murdie, "Blacks in Near-ghettos?: Black Visible Minority Population in Metropolitan Toronto Housing Authority Public Housing Units," *Housing Studies* 9 (1994): 435–457; Eric Fong and Kumiko Shibuya, "The Spatial Separation of the Poor in Canadian Cities," *Demography* 37, no. 4 (November 2000): 449–459.

45. This measure takes size of household into account. For a definition of this variable in the Canadian census, see http://www12.statcan.ca/english/census01/Products/Reference/dict/fam021.htm. Although the measure of low income was invented in the late 1960s, it is first publicly available by census tract in 1981.

46. Gordon M. Fisher, "The Development and History of the Poverty Thresholds," *Social Security Bulletin* 55, no. 4 (1992). For criticism of the poverty line in the United States, see David Brady, "Rethinking the Sociological Measurement of Poverty," *Social Forces* 81, no. 3 (March 2003): 715–752; Patricia Ruggles, *Drawing the Line: Alternative Poverty Measures and their Implications for Public Policy* (Washington, DC: Urban Institute Press, 1990). An additional difference between the two measures complicates this analysis. In the United States "unattached" individuals are combined with individuals living in poor families to create a total count. In Canada the publicly available digital data provide a count of low-income "unattached" individuals and, separately, a count of "economic families" (rather than their individual members) who report low incomes. In the Canadian case, use of two units of measurements—individuals versus families—means that the data cannot be combined to create a figure comparable with that in the United States. I try to compensate for this difference as discussed below.

47. 1980 U.S. Census Extract, Census Tract Data (Philadelphia), NT91A-STF3: "Poverty Status in 1979," National Historical Geographic Information Systems, Minnesota Population Center, University of Minnesota, http://www.nhgis.org (accessed October 2008); Canada Census Analyzer, Census Tract Data, 1981, Toronto, University of Toronto Data Library (accessed October 2008). I arrive at the Toronto figure by adding the proportion of poor economic families in metropolitan Toronto (11.4 percent) together with the percentage of the metropolitan population comprised of poor unattached individuals (4.5). This is only an approximation; the size of the low-income families, relative to other families, is not reported.

48. One tract with a poor majority but only 33 residents was omitted from this analysis.

49. For illustrative purposes, I provide a map of the concentration of economic families who comprised a large majority of the poor in Toronto. However, poor individuals concentrated in similar areas of the city. For detailed maps of both groups, see 1981 Census of Canada, *Metropolitan Atlas Series, Toronto* (Ottawa: Minister of Supply and Services, 1984), 122–129.

50. Vanessa Rosa, "Producing Race, Producing Space: The Geography of Toronto's Regent Park" (M.A. thesis, University of Toronto, 2006); Sean Purdy, "'It Was Tough on Everybody': Low-Income Families and Housing Hardship in Post–World War II Toronto," *Journal of Social History* 37, no. 2 (2003): 457–482, idem, "By the People, for the People: Tenant Organizing in Toronto's Regent Park Housing Project in the 1960s and 1970s," *Journal of Urban History* 30, no. 4 (May 2003): 519–548.

51. Adams et al., "Housing and Neighborhoods," and "Philadephia's Redevelopment Process," in *Philadelphia,* chaps. 3–4.

52. Sociologists and historians detail a prolonged experience of social division in Philadelphia. See William W. Cutler III and Howard Gillette Jr., eds., *The Divided Metropolis: Social and Spatial Dimensions of Philadelphia, 1800–1975* (Westport, CT: Greenwood Press, 1980).

53. Richard Juliani, *Building Little Italy: Philadelphia's Italians before Mass Migration* (University Park: Pennsylvania State University Press, 1998), 230–231.

54. Juliani, *Building Little Italy,* 163.

55. *Report of the Proceeding in the Case of the Contested Election for District Attorney* (Philadelphia: G. T. Stockdale, 1857), 129, 378–379. Cited in Juliani, *Building Little Italy,* 176–179.

56. Richard Juliani, "The Italian Community of Philadelphia," in *Little Italies in North America,* ed. Robert F. Harney and J. Vincenza Scarpaci, 89 (Toronto: Multicultural History Society of Ontario, 1981).

57. Juliani, "Italian Community of Philadelphia," 94–95.

58. Juliani, "Italian Community of Philadelphia," 90; Domenic Vitiello, "National Registry of Historic Places Registration Form: East Passyunk Ave. Historic District" (unpublished Report, 2002).

59. Richard Varbero, "Philadelphia's South Italians in the 1920s," in *The Peoples of Philadelphia: A History of Ethnic Groups and Lower-Class Life, 1790–1940,* ed. Allen Davis and Mark Haller, 266 (Philadelphia: Temple University Press, 1973); Marriage Registry, 1940, St. Thomas Church, St. Thomas Church Archives, Philadelphia, *N* = 100.

60. Marriage Registry, 1950, Annunciation and St. Thomas, Philadelphia. Figures based on sample of fifty couples per church.

61. For contrast to Italian neighborhoods during the period of mass immigration, see Donna R. Gabaccia, *From Sicily to Elizabeth Street: Housing and Social Change among Italian Immigrants, 1880–1930* (Albany: State University of New York Press, 1984); Dino Cinel, *From Italy to San Francisco: The Immigrant Experience* (Stanford, CA: Stanford University Press, 1982); Virginia Yans-McLaughlin, *Family and Community: Italian Immigrants in Buffalo, 1880–1930* (Ithaca: Cornell University Press, 1977); Thomas Kessner, *The Golden Door: Italian and Jewish Immigrant Mobility in New York City, 1880–1915* (New York: Oxford University Press, 1977); Humbert Nelli, *Italians in Chicago, 1880–1930: A Study in Ethnic Mobility* (New York: Oxford University Press, 1970).

62. Minutes, South Philadelphia Committee on Intergroup Affairs, 1960, City of Philadelphia Commission on Human Relations, Houston Community

Centre Papers, URB 40, 19, 124, Urban Archives, Temple University, Philadelphia; Marilyn Zuckerman, "Action of Houston Community Centre during the Racially Tense Weeks of October, 1968," Houston Community Centre Papers, URB 40, 9, 14, Urban Archives, Temple University, Philadelphia; Countryman, *Up South*, 247–255; Stefano Luconi, *From Paesani to White Ethnics: The Italian Experience in Philadelphia* (Albany: State University of New York Press, 2001), 127–128; Kenneth S. Baer, "Whitman: A Study of Race, Class, and Postwar Public Housing Opposition" (senior honors thesis, University of Pennsylvania, 1994). Street violence at later dates was described to me by one of my oral history informants, A. S., interviewed by author, Philadelphia, PA, September 19, 2007.

63. Priests' reports to the archdiocese, 1950, 1990, Annunciation and St. Thomas, Philadelphia Archdiocesan Historical Research Center (PAHRC).

64. This figure includes those who listed "Italian" as a mixed origin. Data accessed at http://www.nhgis.org/ (June 16, 2008). The data in the table include single origins only.

65. Toronto Bureau of Municipal Research, *What Is "The Ward" Going to Do with Toronto?* (Toronto: Bureau of Municipal Research, circa 1918); John Zucchi, *The Italian Immigrants of the St. John's Ward, 1877–1915: Patterns of Settlement and Neighborhood Formation* (Toronto: Multicultural History Society of Ontario, 1980); John Zucchi, *Italians in Toronto: Development of a National Identity, 1875–1935* (Montreal: McGill-Queen's University Press, 1988), chap. 2; Robert F. Harney, "Toronto's Little Italy, 1885–1945," in Harney and Scarpaci, *Little Italies in North America.*

66. Zucchi, *Italians in Toronto,* 41.

67. Zucchi, *Italians in Toronto,* 44; Harney "Toronto's Little Italy," 52.

68. Quoted in Franc Sturino, *Forging the Chain: A Case Study of Italian Migration to North America, 1880–1930* (Toronto: Multicultural Historical Society of Ontario, 1990), 173.

69. Robert F. Harney, "Ethnicity and Neighborhoods," in *Gathering Place: Peoples and Neighbourhoods of Toronto, 1834–1945,* ed. Robert F. Harney, 1–24 (Toronto: Multicultural History Society of Ontario, 1985).

70. Giovanni De Marsico, interview by Robin Healey, Toronto, ON, March and June 1979, Multicultural History Society of Ontario (MHSO), Toronto.

71. Lucy Giovannelli, interview by John Zucchi, Toronto, ON, August 31, 1978, MHSO, Toronto.

72. Zucchi, *Italians in Toronto,* 120–121; Harney "Toronto's Little Italy," 50; Sturino, *Forging the Chain,* 168.

73. Franca Iacovetta, *Such Hardworking People: Italian Immigrants in Postwar Toronto* (Montreal: McGill–Queen's University Press, 1992), chaps. 1 and 2; appendix, tables 4 and 7; Clifford J. Jansen, *Italians in a Multicultural Canada* (Lewistown, NY: Edwin Mellon Press, 1988), 59–62; Knowles, *Strangers at Our Gates,* chap. 8; Kelly and Trebilcock, *Making of the Mosaic,* chap. 8; Historical Statistics of Canada, Section A: Population and Migration, table A385–416, accessed June 19, 2008, http://www.statcan.ca/english/freepub/11-516-XIE/sectiona/A385_416.csv.

74. The notion of a transnational frame of reference expressed in local nodes of activity is well developed in Nicholas DeMaria Harney, *Eh, Paesan! Being Italian in Toronto* (Toronto: University of Toronto Press, 1998)

75. Iacovetta, *Such Hardworking People,* xxii–xxiii, appendix, tables 1, 2, and 5 (204–208); for a contrast to earlier waves of Italian immigrants to Canada, see Robert F. Harney, "Men without Women: Italian Migrants to Canada, 1885–1930," in *The Italian Immigrant Woman in North America,* ed. Betty Boyd Caroli et al. (Toronto: Multicultural History Society of Ontario, 1978), 79–101, idem, "Montreal's King of Italian Labour: A Case Study of Padronism," *Labour/Le Travail* 4 (1979): 57–84; Bruno Ramirez and Michael Del Balso, *The Italians of Montreal: From Sojourning to Settlement, 1900–1921* (Montreal: Associazione di Cultura Populare Italo-Quebecchese, 1980); Bruno Ramirez, *On the Move: French-Canadian and Italian Migrants in the North Atlantic Economy, 1860–1914* (Toronto: McClelland & Stewart, 1991), 93–110; John Zucchi, *The Italian Immigrants of the St. John's Ward, 1877–1915: Patterns of Settlement and Neighborhood Formation* (Toronto: Multicultural History Society of Ontario, 1980), idem, *Italians in Toronto: Development of a National Identity, 1875–1935* (Montreal: McGill–Queen's University Press, 1988); Sturino, *Forging the Chain.*

76. Franca Iacovetta, *Gatekeepers: Reshaping Immigrant Lives in Cold War Canada* (Toronto: Between the Lines, 2006)

77. Jeffry G. Reitz and Raymond Breton, *The Illusion of Difference: Realities of Ethnicity in Canada and the United States* (Ottawa: C. D. Howe Institute publications, 1994). Canadian scholars continue to debate the implications of the multiculturalism policy. See, for example, Reginald Bibby, *Mosaic Madness: The Poverty and Potential of Life in Canada* (Toronto: Stoddart, 1990); Andrew Cardozo and Louis Musto, eds., *The Battle over Multiculturalism* (Ottawa: Pearson-Shoyama Institute, 1997); Will Kymlicka, *Finding Our Way: Rethinking Ethnocultural Relations in Canada* (Toronto: Oxford University Press, 1998); Neil Bissoondath, *Selling Illusions: The Cult of Multiculturalism in Canada,* rev. ed. (Toronto: Penguin Canada, 2002); and Richard J. F. Day, *Multiculturalism in Canada,* 2nd ed. (Toronto: Nelson Thomas Learning, 2002).

78. Meanwhile, significant concentrations of Italians were emerging in the suburbs, particularly in Woodbridge. The emergence of concentrated ethnic suburbs in Toronto, and elsewhere in Canada, deserves further attention elsewhere.

CHAPTER TWO

1. "Public Investigatory Hearings into Discrimination in Housing," box A-2844, group 148, Commission on Human Relations Records, Philadelphia City Archives (PCA), 927, 930.

2. "Public Investigatory Hearings into Discrimination in Housing," box A-2844, group 148, Commission on Human Relations Records, PCA, 931–932.

3. John D., interview by author, Woodbridge, ON, November 23, 2007.

4. James V. Poapst, "Financing of Post-War Housing," in *House, Home, and Community: Progress in Housing Canadians, 1945–1986,* ed. John Miron (Montreal: McGill–Queen's Press, 1993), 95; Robert Murdie, "Residential Mortgage Lending in Metropolitan Toronto: A Case Study of the Resale Market," *Canadian Geographer* 30, no. 2 (1986): 98–100; James V. Poapst, *The Residential Mortgage Market: Working Paper Prepared for the Royal Commission on Banking and Finance* (Ottawa: Queen's Printer, 1962), 3; James V. Poapst, *Developing the Residential Mortgage Market* (Ottawa: Central Mortgage and Housing Corporation, 1975); Philip W. Brown, *Some Perspectives on the Toronto Housing Market: A Review of Recent Studies* (Toronto: Centre for Urban and Community Studies, 1977).

5. The choice of these three parishes is discussed at length in chapter 1. It is worth restating that a single parish—St. Agnes/St. Francis (the parish changed its name and building in 1967)—has always been the sole Italian parish in Toronto's Little Italy, but Italian South Philadelphia has been composed of many parishes, two of which, St. Thomas and Annunciation, are selected for this study.

6. Real Estate Section, *Toronto Star,* May 7, 1960.

7. Philadelphia City Planning Commission, Redevelopment Authority of Philadelphia, and the Philadelphia Housing Authority, *Philadelphia Housing Quality Survey: Passyunk Square Area Report* (Philadelphia: 1950); Domenic Vitiello, "National Registry of Historic Places Registration Form: East Passyunk Ave Historic District" (unpublished report, 2002).

8. City of Toronto Planning and Development Department, "Ward 3 Profile" and "Ward 4 Profile" (Toronto, 1988); David Dunkelman, *Your Guide to Toronto Neighborhoods* (Toronto: Maple Tree Publishing, 1999); Albert Rose, "Prospects for Rehabilitation of Housing in Central Toronto: Report of Research to City of Toronto Planning Board and Central Mortgage and Housing Corporation, 1966," box 82, group 32, City of Toronto Archives (CTA).

9. Toronto Property Tax Assessments, 1950, 1960, 1970, 1980, 1990, CTA; Philadelphia Tax Assessments: 1939, 1951, 1969, Board of Revision of Taxes (BRT) Archive, c/o PCA. Unless otherwise indicated, reports from tax assessments, deeds, and mortgages are derived from sampled streets within the parishes. In each parish, streets were selected on the basis of the addresses of Italian parishioners. Three streets were selected in each parish, and fifty houses on each street were randomly selected for close scrutiny. Figures, therefore, are derived from analysis of 150 properties per parish. On Philadelphia yard space and light, see also, Philadelphia Planning Commission et al., *Philadelphia Housing Quality Survey,* 2–23.

10. Toronto Property Tax Assessments, 1950, 1960, 1970, 1980, 1990, CTA. The census reported that 28 percent of all families in Toronto's Little Italy were "lodgers" in 1951 and 1961: Dominion Bureau of Statistics, *Census of Canada, 1951: Population and Housing Characteristics by Census Tracts, Toronto,* table 2: "Dwelling, household, and family characteristics," tracts 43–45; Dominion Bureau of Statistics, *Census of Canada, 1961, Population and Housing Characteristics by Census Tracts, Toronto,* Bulletin CT-15, table 2, "Household, family, and dwelling characteristics," tracts 43–45. This figure

does not include instances when homeowners created a separate entrance for a self-enclosed apartment.

11. Teo Celotti, interview by John Zucchi, Toronto, ON, July 25, 1977, Multicultural History Society of Ontario, Toronto (MHSO); Daniel Gri, interview by John Zucchi, Toronto, ON, July 5, 1977, MHSO; Rocco Lofranco, interview by Robin Healey, Toronto, ON, June and September 1979, MHSO; Giuseppe Peruzzi, interview by John Zucchi, Toronto, ON, August 10, 1977, MHSO.

12. Angela Iannucci, interview by Doreen Rumack, Toronto, ON, August 9, 1988, Folklife Fonds, Archives of Ontario, Toronto. For similar examples, see Franca Iacovetta, *Such Hardworking People: Italian Immigrants in Postwar Toronto* (Montreal: McGill–Queen's University Press, 1992), 88–90.

13. Vince Pietropaolo, interview by author, Toronto, ON, November 23, 2007. Pietropaolo is referred to in the text as "Vince P." in accordance with the general practice of the book with respect to oral history interviewees, most of whom did not give permission for the use of their full names. However, a public figure and himself a noted observer of Italian life, Pietropaolo has my thanks and explicit credit for his contributions to this book.

14. Paul N., interview by author, Brampton, ON, November 22, 2007.

15. See figure 1.1.

16. Differences between the two data sets (citywide values reported in the census were based on estimates by resident homeowners every ten years, whereas my data for South Philadelphia are drawn from the actual sales of properties) hamper an exact comparison of the timing of price change. To assess the precise timing of price changes citywide, a sample of exchanges would be more useful than census data.

17. Related evidence supports the assumption that most immigrants who attained homeownership would remain in Canada rather than selling property to return to their country of origin. See Richard D. Alba and John R. Logan, "Assimilation and Stratification in the Homeownership Patterns of Racial and Ethnic Groups," *International Migration Review* 26, no. 4 (Winter 1992): 1314–1341. In Toronto, see Thomas Y. Owusu, "To Buy or Not to Buy: Determinants of Home Ownership among Ghanaian Immigrants in Toronto," *Canadian Geographer* 42, no. 1 (Spring 1998): 40–53.

18. See chapter 1.

19. All my work with surnames is based upon a conservative reading of the names in the records in conjunction with catalogs of Italian surnames. Names were deemed Italian only in cases where I could be very confident, as in the case of Antonio Manici. Of course this method is not perfect; Spanish surnames, for example, might be confused with Italian ones. However, this problem is limited because Latin American residents remained a small group in both of my case-study areas, and because given names often help indicate when someone with a Latin-derived last name is likely not Italian (e.g., José). While there remains the possibility that I have included some non-Italians in the Italian group, these are likely more than compensated for by Italian Americans who are left out of the group because they have less obviously Italian names or because their Italian roots are only on the maternal side. Further-

more, the imperfection of the categories tends against most of my findings, which point to the discrepant practices of Italian and non-Italians.

20. The volume of transactions is the weakest point of the real estate data. While deeds can be largely trusted for the information that they contain, calculating the volume of transactions necessarily also draws conclusions from deeds that are absent: the absence of a deed of sale implies the absence of a sale. This logic requires that all existing deeds be unearthed in research, something that cannot be guaranteed. In Philadelphia, where access was granted to summary documents rather than deeds themselves, misplacement of the single summary document resulted in a total loss of information for the property. To minimize this problem, I have used only those properties for which at least one deed of sale was found between 1940 and 1990. While this artificially raises the volume of exchanges (by eliminating those properties never exchanged), it does so in both cities and reduces the special problem of the Philadelphia records. However, by structuring the analysis in this fashion I have generated results that are more useful for comparative conclusions than for an absolute assessment of the volume of transactions in either city.

21. Philadelphia Tax Assessments, 1939, 1951, 1969, BRT archives, c/o PCA.

22. Philadelphia Property Tax Assessments, 1940, 1950, 1960, 1970, BRT archives, c/o PCA; *Cole's Cross Reference Directory, Philadelphia County,* 1980, 1990 (Lincoln, NE: Cole's Publications, 1980, 1990). The sample in 1980 and 1990 includes only those people who could be found in the directory. For 1980 the sample included 150 residents, and for 1990 the number was 109. Figures include resident owners only—renters' names are not available in the Philadelphia tax assessments. Figures include all instances for which the surname of the resident remains identical. Accordingly, they may include the instances of new residents who inherited their house from family members. The difference between the two Philadelphia parishes failed a standard test of statistical significance.

23. C. S., interviewed by author, Philadelphia, PA, September 17, 2007.

24. I. G., interviewed by author, Philadelphia, PA, September 17, 2007.

25. Toronto Property Tax Assessments, 1950, 1960, 1970, 1980, 1990, CTA, $N = 118$.

26. Paul N., interview by author, Brampton, ON, November 22, 2007.

27. Vince Pietropaolo, interview by author, Toronto, ON, November 23, 2007.

28. John D., interview by author, Woodbridge, ON, November 23, 2007; Vince Pietropaolo, interview by author, Toronto, ON, November 23, 2007.

29. The contrast between persistence in Italian South Philadelphia and flux in Toronto's Little Italy suggests that twentieth-century urban life offers intriguing new ground for fined-tuned study of residential mobility akin to the classic work on nineteenth-century cities. While the census manuscripts in both Canada and the United States remain closed to researchers, tax assessments might provide the basis for systematic study. Seminal research on nineteenth urban mobility includes Stephan Thernstrom, *Poverty and Progress: Social Mobility in a Nineteenth-Century City* (Cambridge, MA: Harvard

University Press, 1964), idem, *The Other Bostonians: Poverty and Progress in the American Metropolis* (Cambridge, MA: Harvard University Press, 1973); Peter R. Knights, *The Plain People of Boston, 1830–1860* (New York: Oxford University Press, 1971); Howard P. Chudacoff, *Mobile Americans: Residential and Social Mobility in Omaha, 1880–1920* (New York: Oxford University Press, 1972); Michael B. Katz, *The People of Hamilton, Canada West: Family and Class in a Mid-Nineteenth Century City* (Cambridge, MA: Harvard University Press, 1975); Kathleen Neils Conzen, *Immigrant Milwaukee, 1836–1860: Accommodation and Community in a Frontier City* (Cambridge, MA: Harvard University Press, 1976); Thomas Kessner, *The Golden Door: Italian and Jewish Immigrant Mobility in New York City, 1880–1915* (New York: Oxford University Press, 1977); Theodore Hershberg, ed., *Philadelphia: Work, Space, Family, and Group Experience in the Nineteenth Century* (New York: Oxford University Press, 1981), part 2. For a more recent study that returns to the theme of societies in motion, see Bruno Ramirez, *Crossing the 49th Parallel: Migration from Canada to the United States, 1900–1930* (Ithaca: Cornell University Press, 2001).

30. See Angelo Principe et al., "Italian Canadians, Fascism, and Internment: Black Shirts or Sheep," in *Enemies Within: Italian and Other Internees in Canada and Abroad,* ed. Franca Iacovetta, Robert Perin, and Angelo Principe (Toronto: University of Toronto Press, 2000), part 1.

31. A. S., interviewed by author, Philadelphia, PA, September 19, 2007.

32. C. C., interviewed by author, Philadelphia, PA, September 19, 2007.

33. C. C., interviewed by author, Philadelphia, PA, September 19, 2007.

34. Change over time in this regard failed standard tests of statistical significance. Every effort was made to eliminate transfers of ownership for mere taxation or liability reasons. Hence, transfers between husbands and wives or from a seller to a group that included his/himself were excluded whenever they could be identified. The remaining transactions are more likely to reflect shifts in ownership that carried social meaning within the residential niches; such transfers meant a new occupant or at least a new owner of the property.

35. Violent racial conflict erupted in the housing market under similar conditions elsewhere: Gerald H. Gamm, *Urban Exodus: Why the Jews Left Boston and the Catholics Stayed* (Cambridge, MA: Harvard University Press, 1999), 275–279; John T. McGreevy, *Parish Boundaries: The Catholic Encounter with Race in the Twentieth-Century Urban North* (Chicago: University of Chicago Press, 1996), 19–28; Thomas J. Sugrue, *The Origins of the Urban Crisis: Race and Inequality in Postwar Detroit* (Princeton: Princeton University Press, 1996), 234–258; Arnold R. Hirsch, "Friends, Neighbors, and Rioters," in *Making the Second Ghetto: Race and Housing in Chicago, 1940–1960* (1983; Chicago: University of Chicago Press, 1998), chap. 3; Commission on Human Relations, *Racial Discrimination in Housing: Findings and Recommendations* (Philadelphia: Commission on Human Relations, 1966).

36. Commission on Human Relations, *Racial Discrimination in Housing.*

37. Commission on Human Relations case files, 1959–1965, 1970, 1973–1974, boxes A-2843–46, and A-2848, group 148, Commission on Human Relations Records, PCA.

38. I. G., interviewed by author, Philadelphia, PA, September 17, 2007.

39. A. S., interviewed by author, Philadelphia, PA, September 19, 2007.

40. Bureau of the Census, *1980 Census of Population and Housing, Census Tracts, Philadelphia, PA-NJ Standard Metropolitan Statistical Area,* tracts 28–31.

41. C. C., interviewed by author, Philadelphia, PA, September 19, 2007.

42. The South Philadelphia real estate market speaks to longstanding debates among sociologists and economists about the dynamics of gift exchanges. In South Philadelphia, as I have argued throughout this chapter, social and economic pressures interwove to encourage gift giving. See Marcel Mauss, *The Gift: Forms and Functions of Exchange in Archaic Societies* (1925; New York: Norton, 1967); Claude Lévi-Strauss, *Elementary Structures of Kinship* (London: Eyre & Spottiswoode, 1949); Richard M. Titmuss, *Gift Relationship* (London: George Allen & Unwin, 1970). Thanks to Mathew Creighton for this observation.

43. The difference between Italian and non-Italian gifting in Toronto failed a standard test of statistical significance.

44. John D., interview by author, Woodbridge, ON, November 23, 2007.

45. Toronto Mortgage Records, 1940–1990, Office of Land Title/Land Registry, Ontario Ministry of Consumer Affairs, Toronto, $N = 311$. *Philadelphia Realty Directory and Service* 1940, 1941, 1950, 1951 (Philadelphia), $N = 169$; Philadelphia Mortgages Computer Records, 1980, 1990, Office of Records, City Hall, Philadelphia, $N = 128$. For comparison to analysis that uses larger samples, see Murdie, "Residential Mortgage Lending in Metropolitan Toronto."

46. Poapst, *Residential Mortgage Market,* 3, 13, 34–35.

47. Toronto Mortgage Records, 1940–1990, Office of Land Title/Land Registry, Ontario Ministry of Consumer Affairs, Toronto, $N = 240$.

48. John D., interview by author, Woodbridge, ON, November 23, 2007.

49. See, for example, the real estate sections of the *Corriere Canadese* in the spring and summer of 1975 and 1985. A search the *Toronto Star* real estate section in the first edition of each month in 1950, 1960, 1970, 1980, and 1990 yielded 187 properties listed in Little Italy. As time passed, they were increasingly likely to list with large corporate firms. The circulation of the Saturday edition of the *Toronto Star* ranged between twice and four times that of its nearest competitors. See *Editor & Publisher: International Year Book,* 1940–1990 (New York).

50. "Mights" Metropolitan Toronto City Directory, 1990 (Toronto, 1990).

51. On Portuguese ethnic real estate networks in Toronto in this period, see Carlos Teixeira, "Ethnicity, Housing Search, and the Role of the Real Estate Agent: A Study of Portuguese and Non-Portuguese Real Estate Agents in Toronto," *Professional Geographer* 47, no. 2 (May 1995): 176–84.

52. The role of realtors in segregation is well documented. See, for example, Edward W. Orser, *Blockbusting in Baltimore: The Edmondson Village Story* (Lexington, KY, 1994).

53. Circulation of the Sunday edition of the *Philadelphia Inquirer,* which

carried real estate listings, was always at least 30 percent greater than that of its leading competitor after 1940. See *Editor & Publisher: International Year Book,* 1940–1990 (New York).

54. All figures based on a sum of the first real estate sections of each month in the sample years. In all, forty-five properties listed agents in 1950; eleven were listed with Burton C. Simon.

55. In all, forty-seven properties listed agents in 1970; twenty-three listed with Grasso-Tori.

56. C. C., interviewed by author, Philadelphia, PA, September 19, 2007.

57. "Commission on Human Relations vs. Grasso-Tori Realty Company," August 3 and September 23, 1970, box A-2846, group 148, Commission on Human Relations Records, PCA.

58. Public Investigatory Hearings into Discrimination in Housing, box A-2844, group 148, Commission on Human Relations Records, PCA, 934–35.

59. "Public Investigatory Hearings into Discrimination in Housing," box A-2844, group 148, Commission on Human Relations Records, PCA, 927–32; "Commission on Human Relations vs. Grasso-Tori Realty Company," August 3 and September 23, 1970, box A-2846, group 148, Commission on Human Relations Records, PCA.

60. Lenders' addresses were found by using Philadelphia city directories and phone books to locate the mortgagors listed in *Philadelphia Realty Directory and Service* 1940, 1941, 1950, 1951 (Philadelphia: Philadelphia Real Estate Directory) and Philadelphia Mortgages Computer Records, 1980, 1990, Office of Records, City Hall. On the origins of savings and loan associations in Philadelphia, see William N. Loucks, *The Philadelphia Plan of Home Financing: A Study of the Second Mortgage Lending of Philadelphia Building and Loan Associations* (Chicago: Institute for Research in Land Economics, 1929).

CHAPTER THREE

1. For a detailed description of the parish locations, see chapter 1.

2. Gerald H. Gamm, *Urban Exodus: Why the Jews Left Boston and the Catholics Stayed* (Cambridge, MA: Harvard University Press, 1999); John T. McGreevy, *Parish Boundaries: The Catholic Encounter with Race in the Twentieth-Century Urban North* (Chicago: University of Chicago Press, 1996), 19–28; Jordan Stanger-Ross, "Neither Fight nor Flight: Urban Synagogues in Postwar Philadelphia," *Journal of Urban History* 32, no. 6 (September 2006): 791–812.

3. Letter from Rev. Vincent Mele to Cardinal McGuigan, August 21, 1958, General Correspondence, 1915–1961, St. Agnes Parish Papers, Archives of the Roman Catholic Archdiocese of Toronto (ARCAT).

4. Letter from Adolph Baldolini to Rev. Philip Pocock, October 3, 1962, St. Agnes Parish Papers, ARCAT.

5. "St. Agnes Chronicle," December 24, 1964, St. Francis Parish Archives, Toronto.

6. Letter from Chancellor Thomas B. Fulton to Rev. Matthew DeBenedic-

tis, July 14, 1967, Parish History: St. Agnes and St. Francis Unite, 1967–1968, St. Agnes Parish Papers, ARCAT.

7. Spiritual Statistics for the years ending December 31, 1970, and 1978, Parish Spiritual Statistics, St. Francis of Assisi, 1914–1978, Spiritual Statistics, St. Francis of Assisi, ARCAT; "Forming 'Community' Enlivens Italian Parish," *Catholic Register* (Toronto) March 30, 1991. The spiritual statistics for St. Agnes/St. Francis are unfortunately very uneven; participation is reported only in the 1970s.

8. Annunciation parishioner, group interview by author, Philadelphia, PA, May 10, 2004.

9. Annunciation group interview.

10. Spiritual Statistics, 1980, Annunciation and St. Thomas, Philadelphia Archdiocesan Historical Research Center (PAHRC).

11. On the diverse roles of churches in immigrant communities in postwar Toronto, see Roberto Perin, "Churches and Immigrant Integration in Toronto, 1947–65," in *The Churches and Social Order in Nineteenth- and Twentieth-Century Canada,* ed. Michael Gauvreau and Ollivier Hubert (Montreal: McGill–Queen's University Press, 2006), 274–92. For a portrayal of the connection between the church and community in Catholic neighborhoods in the postwar United States, see Alan Ehrenhalt, *The Lost City: Discovering the Forgotten Virtues of Community in Chicago of the 1950s* (New York: Basic Books, 1995), chap. 5.

12. Letter from Italian Canadian parishioner [name suppressed on request of the archive] to Rev. Matthew M. D. Benedictis, September 14, 1970, Parish Community: Parishioners petition for the return of Fr. Ambrose De Luca, St. Francis Parish Papers, ARCAT.

13. Petition to the archbishop, September 1, 1970, Parish Community: Parishioners petition for the return of Fr. Ambrose De Luca, St. Francis Parish Papers, ARCAT; letter from Thomas Fulton to Father Matthew DeBenedictis, September 7, 1970, Parish Community: Parishioners petition for the return of Fr. Ambrose De Luca, St. Francis Parish Papers, ARCAT.

14. Letter from Italian Canadian parishioner [name suppressed on request of the Archive] to Archbishop Charles McGuigan, May 1, 1935, General Correspondence, 1915–1961, St. Agnes Parish Papers, ARCAT.

15. Riccardo Politicchia, interview by John Zucchi, Toronto, ON, December 15, 1977, Multicultural History Society of Ontario (MHSO). Father Riccardo was priest from 1936 to 1943 and again from 1944 to 1958; he was associate priest from 1961 to 1969. See Parish Appointments, Parish History, St. Agnes Parish Papers, ARCAT.

16. Father Gregory Botte, interview by author, Toronto, ON, March 22, 2004; Ezio Marchetto, "Padre Riccardo: Pioneer in the Italian Community of Toronto," *Italian Canadian,* vol. 4 (Toronto, 1988).

17. Ottorino Bressan, interview by John Zucchi, Toronto, ON, June 22, 1977, MHSO; Giovanni De Marsico, interview by Robin Healey, Toronto, ON, March and June 1979, MHSO; Lucy Giovannelli, interview by John Zucchi, Toronto, ON, August 31, 1978, MHSO; Giuseppe Peruzzi, interview by

John Zucchi, Toronto, ON, August, 10 1977, MHSO; Peter Bossa, interview by John Zucchi, Toronto, ON, August 22, 1977, MHSO.

18. Letter from Rev. George Nincheri to Archbishop Philip Pocock, May 23, 1964, General Correspondence, 1962–1984, St. Francis Parish Papers, ARCAT.

19. Raffle flyer, June 17, 1984, General Correspondence, 1962–1984, St. Francis Parish Papers, ARCAT.

20. Evidence of these activities abounds in parish weekly and monthly bulletins and calendars. See St. Thomas Parish Calendars, 1940–1942, 1943–1955, 1976–1986, PAHRC; Annunciation Parish Calendars, 1948–1990, Annunciation Parish Archives, Philadelphia.

21. Spiritual Statistics, 1950, 1970, Annunciation and St. Thomas, PAHRC. The CYO accounted for 2,350 of the total association members in 1970.

22. Annunciation group interview.

23. St. Thomas Parish Calendar, September 26, 1976, p. 2, PAHRC. The calendar regularly reported income from bingo in its 1976 issues.

24. The carnival occurred every summer in June. See, for example, Annunciation Parish Calendar, April-May 1949, p. 2, April 20, 1968, p. 17, May 27, 1972, p. 1, June 30, 1984, p. 3–6, Annunciation Parish Archives, Philadelphia.

25. Richard A. Varbero, "Philadelphia's South Italians in the 1920s," in *The Peoples of Philadelphia: A History of Ethnic Groups and Lower-Class Life, 1790s–1940,* ed. Allen F. Davis and Mark H. Haller (Philadelphia: Temple University Press, 1973), 255–275; Thomas Kessner, *The Golden Door: Italian and Jewish Immigrant Mobility in New York City, 1880–1915* (New York: Oxford University Press, 1977), 95–99; and McGreevy, *Parish Boundaries,* 21, 237.

26. Annunciation Parish Calendars, April–May 1948, Annunciation Parish Archives, Philadelphia.

27. Spiritual Statistics, 1970, St. Thomas and Annunciation Parish, PAHRC. Rising Catholic school enrollments in South Philadelphia fit into broader trends in Philadelphia, where, by the mid-1960s, some 60 percent of the city's white high school students were in Catholic schools (McGreevy, *Parish Boundaries,* 237).

28. Robert M. Stamp, *The Schools of Ontario, 1876–1976* (Toronto: University of Toronto Press, 1982), 26.

29. James S. Brown, "The Formation of the Metropolitan Separate School Board (Toronto), 1953–1978," and Franklin Walker, "Who Ran the Catholic Schools? The Role of Clergy, Trustees and the Provincial Government in the Separate School System," in *The History of Catholic Education in Ontario: A Reader,* ed. Robert Dixon and Mark G. McGowan, 237–246, 279–298 (Toronto: Canadian Scholars' Press, 1997).

30. Memo, "Rev. Adolph Baldolini Visits the Chancery Office to Discuss Certain Problems Relative to St. Agnes Parish," September 6, 1963, General Correspondence, 1963–1968, St. Agnes Parish Papers, ARCAT.

31. Spiritual Statistics for the years ending December 31, 1970 Parish

Spiritual Statistics, St. Francis of Assisi, 1914–1978, Spiritual Statistics, St. Francis of Assisi, ARCAT. On rising enrollment, see Brown, "Formation of the Metropolitan Separate School Board," 240–242.

32. Robert A. Orsi, *The Madonna of 115th Street: Faith and Community in Italian Harlem, 1880–1950,* 2nd ed. (New Haven, CT: Yale University Press, 2002); Robert A. Orsi, ed., *Gods of the City: Religion and the American Urban Landscape* (Bloomington: Indiana University Press, 1999). For a discussion of street processions at an earlier juncture in Philadelphia history, see Susan Davis, *Parades and Power: Street Theatre in Nineteenth-Century Philadelphia* (Berkeley: University of California Press, 1988).

33. See Robert A. Orsi, "The Religious Boundaries of an Inbetween People: Street *Feste* and the Problem of the Dark-Skinned Other in Italian Harlem, 1920–1990," *American Quarterly* 44, no. 3 (1992): 313–347; Joseph P. Sciorra, "'We Go Where the Italians Live': Religious Processions as Ethnic and Territorial Markers in Multi-Ethnic Brooklyn Neighborhood," in Orsi, *Gods of the City,* 310–340. Nicholas Harney argues that the same was true in Toronto, although, as I will argue below, I believe that processions worked differently in Toronto's Little Italy than in American sites such as Philadelphia and New York. See Nicholas Demaria Harney, "Ethnicity, Social Organization, and Urban Space: A Comparison of Italians in Toronto and Montreal," in *Urban Enigmas: Montreal, Toronto, and the Problem of Comparing Cities,* ed. Johanne Sloan (Montreal: McGill–Queen's University Press, 2007), 185–190.

34. St. Thomas school teacher, interview by author, Philadelphia, PA, May 10, 2004; Annunciation school teacher, interview by author, Philadelphia, PA, May 11, 2004.

35. Annunciation group interview.

36. Father Arthur Taraborelli interview by author, Philadelphia, PA, May 10, 2004.

37. Annunciation group interview. Large Catholic, and indeed Italian processions and festivals continue to exist in the United States. Historians might examine why, for example, the Giglio Festival has persisted on such a large scale in Williamsburg, Brooklyn, but no event of a similar scale persisted in South Philadelphia parishes.

38. Annunciation Parish Calendars, October 1948, p. 4, August 1953, p. 1, Annunciation Parish Archives, Philadelphia.

39. Annunciation Parish Calendar, September 1949, p. 4, Annunciation Parish Archives, Philadelphia.

40. Annunciation Parish Calendar, November 1950, p. 3, September–October 1951, p. 3, Annunciation Parish Archives, Philadelphia.

41. Spiritual Statistics, 1950, Annunciation Parish, PAHRC.

42. Annunciation Parish Calendars, September 1952, p. 1, November 1957, p. 16, Annunciation Parish Archives, Philadelphia.

43. Annunciation Parish Calendar, August 1953, p. 1, Annunciation Parish Archives, Philadelphia. On the origins of the procession as anticommunist, see Annunciation Parish Calendar October 1949, p. 1, Annunciation Parish Ar-

chives, Philadelphia. On the wider role of anticommunism in twentieth-century Catholicism, see McGreevy, *Parish Boundaries,* 64–66, 105–106.

44. Annunciation Parish Calendar November 1952, p. 5, Annunciation Parish Archives, Philadelphia.

45. Annunciation Parish Calendar, September 1949, p. 4, November 1957, p. 16, Annunciation Parish Archives, Philadelphia.

46. Annunciation Parish Calendar, November 1950, p. 3, Annunciation Parish Archives, Philadelphia.

47. For the most part, these outsiders were also Italian South Philadelphians. St. Thomas Parish lay to the west of Annunciation and to the north were areas only slightly less heavily Italian than Annunciation. However, in the early postwar years parish boundaries continued to carry social significance even within Italian South Philadelphia, as I will argue in chapter 4. By the time the procession waned, parish boundaries were becoming less significant.

48. The Annunciation Holy Name Society declined from 629 to 175 members between 1950 and 1970. See St. Thomas Spiritual Statistics, 1950, 1970, PAHRC.

49. Annunciation Parish Calendar, September 1972, p. 9, Annunciation Parish Archives, Philadelphia.

50. St. Thomas Parish Calendar, 24 July 1977, p. 3, PAHRC.

51. St. Thomas Parish Calendar, July 24, 1977, p. 3; August 7, 1977, p. 3; August 14, 1977, p. 3; October 2, 1977, p. 3, PAHRC.

52. St. Thomas Parish Calendar, August 28, 1977, p. 3, PAHRC. On Frank Rizzo, see Stefano Luconi, "Frank L. Rizzo and the Whitening of Italian Americans in Philadelphia," in *Are Italians White? How Race Is Made in America,* ed. Jennifer Guglielmo and Salvatore Salerno, 177–191 (New York: Routledge, 2003); Stefano Luconi, "From Italian Americans to White Ethnics," in *From Paesani to White Ethnics: The Italian Experience in Philadelphia* (Albany: State University Press of New York, 2001), chap. 6; Joseph R. Daughen and Peter Binzen, *The Cop Who Would Be King: Mayor Frank Rizzo* (Boston: Little Brown, 1977); Fred Hamilton, *Rizzo: From Cop to Mayor of Philadelphia* (New York: Viking, 1973).

53. St. Thomas Parish Calendar, October 9, 1977, p. 3, PAHRC.

54. St. Thomas Parish Calendars, October 8, 1978, p. 3; September 7, 1980, p.3; October 4, 1981, p. 3; July 11, 1982, p. 3; August 1, 1982, p. 3; July 31, 1983, p. 3; October 7, 1984, p. 3; September 1, 1985, p. 3; August 31, 1986, p. 3, PAHRC.

55. St. Thomas Spiritual Statistics, 1990, PAHRC.

56. Photographs courtesy of Father Taraborelli, St. Thomas Parish; procession route in St. Thomas Parish Calendar, July 29, 1984, p. 3, PAHRC. Census tract 36-B, a majority black tract, began at Eighteenth Street.

57. Photographs courtesy of Father Taraborelli.

58. Father Taraborelli interview.

59. St. Thomas school teacher interview; St. Thomas Parish Calendar, August 22, 1982, PAHRC.

60. St. Thomas Parish Calendar, August 28, 1983, PAHRC. For a broader issue of race and Catholic schools during this period, see McGreevy, *Parish Boundaries*, 234–247.

61. McGreevy, *Parish Boundaries.*

62. Father Taraborelli interview.

63. See McGreevy, *Parish Boundaries;* Robert A. Orsi, review of McMahon, *What Parish Are You From?* and McGreevy, *Parish Boundaries* in *American Historical Review* 101, no. 5 (December 1996): 1640–1641.

64. Father Taraborelli interview. These precise events could not be corroborated in any other source that I consulted. However, the priest's remarks reflect a sense of threat on parish streets, and this sense was palpable in all of the St. Thomas materials. A closer analysis of the *Festa* itself—perhaps rooted in further oral historical research (among students and teachers at the school)—might yield greater insight into the fascinating and paradoxical history of the *Festa Italiana* in St. Thomas Parish.

65. Letter from Riccardo Politicchia to John Harris, May 29, 1937, General Correspondence, 1915–1961, St. Agnes Parish Papers, Archives of the Roman Catholic Archdiocese of Toronto (ARCAT).

66. Paul N., interview by author, Brampton, ON, November 22, 2007.

67. This description compiled from a number of articles in the *Corriere Canadese* (Toronto) between the 1950s and the 1970s. See: "A S. Agnese degna celebrazione di S. Antonio di Padova," June, 21 1955; Photographs and caption, June 24, 1958; "La festa solenne in onore di S. Antonio," June 21, 1961; "Gli Italiani celebrano la festa di Sant'Antonio," June 17, 1967; "S. Antonio festeggiato nella 'Little Italy,'" June 17, 1969; "Solemni festiggiamenti in onore di S. Antonio organizzati dalla parrocchia di S Francesco d'Assisi," June 16–17, 1975; "Una difesa quasi multiculturale delle feste dei santi all'Italiana," in *Il Samaritano* (Sunday insert to the paper), June 18, 1978.

68. Author's translation. Photographs and caption, *Corriere Canadese,* June 24, 1958; "La festa solenne in onore di S. Antonio," *Corriere Canadese,* June 21, 1961.

69. "Gli Italiani celebrano la festa di Sant'Antonio," *Corriere Canadese,* June 17, 1967.

70. "Solemni festiggiamenti in onore di S. Antonio organizzati dalla parrocchia di S Francesco d'Assisi," *Corriere Canadese,* June 16–17, 1975.

71. "St. Agnes Chronicle," June 13, 1965, June 13, 1966, May 20, 1966, June 16, 1967, St. Francis Parish Archives, Toronto.

72. "St. Anthony Feast 1983," Parade Processions, 1976–1983, Chancellor in Spirtualibus Fonds, ARCAT. The account is not credited to any specific parish, but no other parish held an event of this magnitude for St. Anthony, so it is safe to assume that it was for St. Francis Church.

73. Letters of application from priests to archdiocese, CU 21.01-CU21.29, Archbishop Philip Pocock Fonds, ARCAT.

74. Parish Procession Applications, 1986, folder: Parades-Processions, 1985–1986, Permits, Lists, Letters, box: Parade Processions, 1976–1986,

Chancellor in Spirtualibus Fonds, ARCAT. For a description of the smaller Italian festivals and processions in Toronto, see Bruce B. Giuliano, *Sacro o Profano? A Consideration of Four Italian-Canadian Religious Festivals* (Ottawa: National Museums of Canada, 1976). The CHIN International Picnic, initiated by Little Italy notable Johnny Lombardi, drew more participants than the Good Friday procession. However, as a secular multiethnic affair held outside of Little Italy, it does not fall under consideration here. For comparative discussion of the two events, see Nicholas Harney, "Ethnicity, Social Organization, and Urban Space."

75. See Robert Orsi, "Everyday Miracles: The Study of Lived Religion," in *Lived Religion in America: Toward a History of Practice,* ed. David D. Hall, 3–21 (Princeton, NJ: Princeton University Press, 1997).

76. Paul N., interview by author, Brampton, ON, November 22, 2007.

77. "Application for Permission to Hold a Parade," St. Francis, 1986, folder: Parades-Processions, 1985–1986, Permits, Lists, Letters, box: Parade-Processions, 1976–1986, Chancellor in Spirualibus Fonds, ARCAT.

78. "Little Italy's Day for Our Lord, Jesus," *Catholic Register* (Toronto), April 25–May 1, 1987.

79. "Le processioni del Venerdi Santo seguite da almeno 60,000 persone," *Corriere Canadese,* April 5–6, 1983.

80. Francesca S., interview by author, Scaborough, ON, November 22, 2007.

81. Paul N., interview by author, Brampton, ON, November 22, 2007.

82. Vince Pietropaolo, interview by author, Toronto, ON, November 23, 2007.

83. Letter from Rev. Matthew De Benedictis to Rev. Thomas B. Fulton, Parish History: Transfer to Portuguese congregation, 1970, St. Agnes Parish Papers, ARCAT.

84. Parades-Processions, 1985–1986, Permits, Lists, Letters, Box: Parade Processions, 1976–1986, Chancellor in Spirtualibus Fonds, ARCAT.

85. The following analysis is based on six separate sources. Four volumes represent the two South Philadelphia parishes: for Annunciation Parish, I use a memorial volume produced in 1937—the only such volume available for the parish—and a monthly calendar from 1962; for St. Thomas, I select the parish's "Diamond Jubilee" publication in 1960 and its centennial volume of 1985. In Toronto's Little Italy, the best volumes are the parish commemorative publications for the twenty-fifth anniversary of the arrival of the Franciscan Friars to the clergy in 1959, and the seventy-fifth anniversary of the parish in 1978. See "The Annunciation Church, Philadelphia Pa., 1860–1937," box 4, group 102, Parish Histories, PAHRC; "Saint Thomas Aquinas Parish Diamond Jubilee," 1960, box 48, group 102, Parish Histories, PAHRC; Annunciation Monthly Calendar, January 13, 1962, Annunciation Parish Archives, Philadelphia; "St. Thomas Aquinas Parish, 1885–1985," box 48, group 102, Parish Histories, PAHRC; "Silver Jubilee Commemoration at Saint Agnes Church, Grace and Dundas Streets Toronto, Canada," Parish History: Franciscan's Twenty-fifth Anniversary Publication, 1959, St. Agnes Parish Papers, ARCAT; "Diamond Jubilee Celebration, St. Francis/Old St. Agnes Parish

Community, 1903–1978," Publications: Seventy-fifth Anniversary Book, St. Francis Parish Papers, ARCAT.

In all, the volumes provide four separate indicators of the geography of business support for the parishes in South Philadelphia, and two snapshots of the geography of church support in Toronto. While these sources vary from one another, they provide a basis for comparison, and dramatic differences between the cities emerge in analysis of the volumes. The consistency of these differences suggests that they indeed capture the divergent social geographies of parish life in the two enclaves.

86. Only those ads that gave a personal name, 68 percent of the total, could be used in this analysis. The total counts in these analyses by year: Philadelphia, 1937, N=66, 1960, N = 91, 1962, N = 21, 1985, N = 35; Toronto, 1959, N = 86, 1978, N = 65.

87. Only those ads that gave a personal name could be used in this analysis. The total counts in these analyses by year 1959, N = 86, 1978, N = 65.

88. Distances mentioned in this chapter are calculated using a geographic information system (GIS) to find coordinates for addresses in Philadelphia and Toronto and the following formula to calculate the distance between two selected points:

$$\text{Distance in miles} = \text{sqrt}(x^2 + y^2)$$
$$\text{where } x = 69.1 * (\text{lat2}—\text{lat1})$$
$$\text{and } y = 69.1 * (\text{lon2}—\text{lon1}) * \cos(\text{lat1}/57.3)$$

89. In 1959, 40 out of 121 businesses whose locations could be identified were within the parish, and in 1978, 36 of 107 businesses were in the parish.

CHAPTER FOUR

1. A. S., interviewed by author, Philadelphia, PA, September 19, 2007.

2. Annunciation parishioners, group interview by author, Philadelphia, PA, May 10, 2004.

3. In Canada, ongoing Italian immigration kept Italian endogamy rates higher than in the postwar United States, but neither nation saw numbers as high as those in the parishes studied here. On the broader patterns in each country, see Richard Alba, *Italian Americans: Into the Twilight of Ethnicity* (Englewood Cliffs, NJ: Prentice-Hall, 1985); Stanley Lieberson and Mary C. Waters, *From Many Strands: Ethnic and Racial Groups in Contemporary America* (New York: Russell Sage Foundation, 1988); Madeline Richard, *Ethnic Groups and Marital Choices: Ethnic History and Marital Assimilation in Canada, 1871 and 1971* (Vancouver: UBC Press, 1991); Jeffrey G. Reitz and Raymond Breton, *The Illusion of Difference: Realities of Ethnicity in Canada and the United States* (Ottawa: CD Howe Institute, 1994). The analysis that follows is based on Catholic marriage registries in the parishes at the core of this study, rather than the larger survey or census data that are often used in this kind of analysis. The registries, of course, are not directly comparable with census figures. However, the marriage registries provide insight into people

who were enveloped by the institutions and networks that animated Italian ethnicity—the kind of people likely to live in densely Italian neighborhoods. People of Italian birth or extraction who found companionship elsewhere and built their lives in different institutions and different social surroundings, are a somewhat separate population.

4. Lucy Giovannelli, interview by John Zucchi, Toronto, ON, August 31, 1978, Multicultural History Society of Ontario (MHSO).

5. I. G., interviewed by author, Philadelphia, PA, September 17, 2007.

6. All figures in this chapter based on analysis of a random selection of at least 50 brides per parish in each sample year. In cases when fewer than 50 marriages occurred in a given sample year, all marriages in that year were recorded and then supplemented by a sample from the following year.

7. The marriage registries of local churches provide only a partial picture of the youth social networks in each area. Since Catholics normally marry in the woman's home parish, the records from Annunciation, St. Thomas, and St. Agnes/St. Francis especially illuminate the social experiences of local women. Young men marrying Catholic women from other places seldom left a trace in their own parish registries. Similarly, marriage records underrepresent unions between Italian Catholics and members of other religious faiths. Although marriages between parishioners and non-Catholics appear in the records of all three churches in this study, many of those who married non-Catholics likely took their vows at their partner's church or in civil ceremonies. These two limits of the registries run together as Italian men were more likely than their female peers, for most of the twentieth century, to marry non-Italians. Marriage registries cannot, therefore, be taken to represent the experiences of the missing men.

8. In 1960, almost two-thirds of the Italian-born brides were from Calabria. They married Calabrese grooms 77 percent of the time.

9. Francesca S., interview by author, Scaborough, ON, November 22, 2007.

10. Francesca S., interview by author, Scaborough, ON, November 22, 2007. Foligno is located significantly north of Sicily in the central Italian region of Umbria.

11. See John Zucchi, *Italians in Toronto: Development of a National Identity, 1875–1935* (Montreal: McGill–Queen's University Press, 1988).

12. Vince Pietropaolo, interview by author, Toronto, ON, November 23, 2007.

13. Zucchi, *Italians in Toronto.*

14. The figures that follow are based upon addresses recorded for brides and grooms in the marriage registries. I was somewhat surprised to find only a negligible increase in the reported instances of cohabitation before marriage. The absence of an increase, despite the well-documented increase in premarital cohabitation in this period, likely reflects the attempt of some young couples to avoid conflict with priests and church officials. In response to my question in this regard, a bookkeeper at one of the parishes answered sharply, "Well, they *better* not be living together before they get married!" Nonetheless, even

if young couples adjusted addresses for the sake of propriety, they probably did not invent the addresses. Instead, such couples probably offered either a parent's address or a previous address. Accordingly, address information remains relevant, even when it is not perfectly accurate. The act of cohabitation, even if it preceded marriage, likely connected the addresses provided by a bride and groom.

15. C. S., interviewed by author, Philadelphia, PA, September 17, 2007.

16. A. S., interviewed by author, Philadelphia, PA, September 19, 2007.

17. Annunciation parishioners, group interview by author, Philadelphia, PA, May 10, 2004.

18. A. S., interviewed by author, Philadelphia, PA, September 19, 2007.

19. Audrey Geniole, interview by A. McPeek, Toronto, ON, September 20, 1977, MHSO.

20. Rocco Lofranco, interviewed by Robin Healey, Toronto, ON, June–September 1979, MHSO. Pisticci is in the southern Italian region of Basilicata.

21. Racco family, interview by Jocelyn Paquette, Toronto, ON, February 15, 1988, MHSO.

22. Paul N., interview by author, Brampton, ON, November 22, 2007.

23. Vince Pietropaolo, interview by author, Toronto, ON, November 23, 2007.

24. Marriage Registry, 1950, Annunciation and St. Thomas, Philadelphia. Figures based on sample of 50 brides per church.

25. Marriage Registry, 1950 and 1980, Annunciation and St. Thomas, Philadelphia. The difference between the two churches failed a standard test of statistical significance.

26. C. C., interviewed by author, Philadelphia, PA, September 19, 2007.

27. A. S., interviewed by author, Philadelphia, PA, September 19, 2007.

28. I. G., interviewed by author, Philadelphia, PA, September 17, 2007.

29. For the purposes of this analysis, I have used the larger area of Italian settlement bounded in the south by Oregon Avenue, the north by South Street, the east by Sixth Street, and the west by Twenty-third Street.

30. University of Pennsylvania Library, *1950 Philadelphia Census Tract Data,* extracted from "The Census Tract Data, 1950: Elizabeth Mullen Bogue File" (National Archives and Records Administration), http://data.library.upenn.edu/phila.html (accessed on March 11, 2005); U.S. Census Bureau, *1980 Census of Population and Housing. Census tracts, Philadelphia, PA-NJ Standard Metropolitan Statistical Area,* table P-1, "General Characteristics of Persons" (Washington, DC: Bureau of the Census, 1983).

31. The difference between the two parishes in this regard failed a standard test of statistical significance.

32. Paul N., interview by author, Brampton, ON, November 22, 2007.

33. Vince Pietropaolo, interview by author, Toronto, ON, November 23, 2007.

34. Sonia Cancian, "What Do Love Letters Tell Us about Italian Postwar Migration to Canada," paper presented at the Canadian Society for Italian Studies Conference, Vancouver 2008.

35. Paul N., interview by author, Brampton, ON, November 22, 2007.

36. Antoniette Ranieri, interview by Giocchio Di Nardo, Toronto, ON, March 23, 1980, MHSO.

37. Vince Pietropaolo, interview by author, Toronto, ON, November 23, 2007.

38. Francesca S., interview by author, Scaborough, ON, November 22, 2007.

39. Frank Stella, interview by Carole Carpenter, Toronto, ON, February 24, 1988, Folklife Fonds, Archives of Ontario, Toronto.

40. Vittorio Zavagno, interviewed by John Zucchi, Weston, ON, 7 July 1977.

41. Mirella Borsoi, interviewed by Doreen Rumack, Toronto, ON, March 24, 1988, Folklife Fonds, Archives of Ontario, Toronto.

42. Lucy Giovannelli, interviewed by John Zucchi, Toronto, ON, August 31, 1978, MHSO.

43. This claim would be much strengthened by direct comparison of young men and women, which is not possible in the present analysis because the marriage rolls selected provide less reliable information on local men. Fuller analysis of the role of gender in the geography of romance could be accomplished with a wider sampling of Catholic parishes or use of the municipal marriage rolls.

44. C. S., interviewed by author, Philadelphia, PA, September 17, 2007.

CHAPTER FIVE

1. Vince Pietropaolo, interview by author, Toronto, ON, November 23, 2007.

2. Vincenzo Pietropaolo, *Not Paved with Gold: Italian-Canadian Immigrants in the 1970s* (Toronto: Between the Lines, 2006), 2.

3. Pietropaolo, *Not Paved with Gold,* 3; Vince Pietropaolo, interview by author, Toronto, ON, November 23, 2007.

4. Judge Harry Waisberg, *Report of the Royal Commission on Certain Sectors of the Building Industry,* V1 (Toronto: Queen's Printer of Ontario, 1974); Franca Iacovetta, *Such Hardworking People: Italian Immigrants in Postwar Toronto* (Montreal: McGill–Queen's University Press, 1992), 156–162.

5. Iacovetta, "Men, Work, and the Family Economy," in *Such Hardworking People,* chap. 3.

6. Italian construction in Toronto was not without its controversies. Reports of graft and violence in the industry motivated a Royal Commission inquiry. The construction industry played host to shady characters and illicit deals, to be sure, but the Royal Commission on this topic exposed not only illegality but also the importance of ethnic social networks to a crucial economic sector. See Ontario, Royal Commission on Certain Sectors of the Building Industry, *Transcripts of Public Hearings–Royal Commission on Certain Sectors of the Building Industry* (Toronto: Legislative Assembly of Ontario Library). For labor politics in the same years, see Franca Iacovetta, *Such Hardworking People: Italian Immigrants in Postwar Toronto* (Montreal: McGill-Queen's University Press, 1992).

7. My argument in this chapter relies primarily on censuses, which provide systematic information on labor choices and the movement of workers through city space. The relevant data come in two forms. The first provides detailed information on workforce participation. In the United States, public use microdata samples provide information about individuals, who can then be considered collectively. In Canada, where the public use samples include a less-useful industry variable, special tabulations of the census provide the aggregated industrial occupations of the entire Italian origins population in metropolitan Toronto. Unfortunately, these data are not publicly available and cannot, by law, be made so by data users who purchase only limited access to the files. For the analysis, I use the special tabulations of the 1951 and 1961 censuses graciously provided for my use by Franca Iacovetta (who used them in *Such Hardworking People,* see appendix, tables 11–14), and my own purchased tables for 1971 and 1981. These sources provide detailed information about workforce participation, but they lack geographic specificity. Such data permit analysis of how ethnicity shaped labor choices on a metropolitan scale, but they cannot be used to identify the residents of the particular neighborhoods at the center of this study (while the Canadian census can provide tables that include greater geographic detail in the census years I requested, the price quote—in excess of $30,000—made the analysis unfeasible). The second data source provides more geographically specific information by detailing aspects of workforce activity by census tract, a smaller geographic unit within the city. The geographic specificity of census tracts permits analysis of Toronto's Little Italy and Italian South Philadelphia in particular. Such data, however, have two important limitations: they fail to distinguish between Italian and non-Italian residents of the tracts, a problem that is especially important in Toronto, where Little Italy remained ethnically heterogeneous, and they provide less detailed information about the workforce. In this chapter, I use these two imperfect sources to describe the role of Italian ethnicity within each city's labor market and then to examine the social geography of Italian workforce participation.

8. Roger Waldinger, *Still the Promised City? African Americans and New Immigrants in New York* (Cambridge, MA: Harvard University Press, 1996), 20–32. See also Suzanne Model, "The Ethnic Niche and the Structure of Opportunity: Immigrants and Minorities in New York," in *The "Underclass" Debate: Views from History,* ed. Michael Katz, 161–193 (Princeton: Princeton University Press, 1993); Mark Granovetter, *Getting a Job,* 2nd ed. (Chicago: University of Chicago Press, 1995); Jeffrey G. Reitz, *Ethnic Group Control of Jobs,* Centre for Urban and Community Studies, research paper no. 133 (Toronto: University of Toronto, 1982).

9. Model, "Ethnic Niche," 164. I follow Waldinger in applying this definition to the industry variable of the census (IND1950 in IPUMS), and limiting analysis to industries employing over 1,000 people. For analysis of female-only niches, I dropped the threshold to 500 workers. See Waldinger, *Still the Promised City?* 59, 340 note 3.

10. Unless otherwise specified, all data reported in the pages that follow derive from the following sources: U.S. microdata: Steven Ruggles et al.,

Integrated Public Use Microdata Series: Version 3.0 (Minneapolis: Minnesota Population Center, 2004): 1950 General Sample (1 percent), 1970 Form 2 Metro Sample (1 percent), 1970 Form 2 Neighborhood Sample (1 percent); 1970 Form 2 State Sample (1 percent), 1980 Metro Sample (1 percent), 1980 State Sample (5 percent), 1990 Metro Sample (1 percent), 1990 State Sample (5 percent). U.S. tract-level data: University of Pennsylvania Library, *Philadelphia Census Tract Data and Maps, 1940–2000,* http://data.library.upenn.edu/phila.html (accessed on March 11, 2005). Canadian data: University of Toronto Data Library Service, *Census of Canada: Public Use Microdata Files 1871–1996,* http://www.chass.utoronto.ca.myaccess.library.utoronto.ca/datalib/major/canpumf.htm#1971 (accessed March 25, 2005), "Individual Files," 1971 (1 percent), 1981 (2 percent), 1991 (3 percent); and special tabulations of the census (industry by ethnic origins) for 1951, 1961, 1971, 1981. Canadian tract level-data: University of Toronto Data Library Service, Census of Canada, 1961, 1971, 1981, 1991, Basic Summary Tabulations, Industry by Tract, http://www.chass.utoronto.ca.myaccess.library.utoronto.ca/datalib/major/major.htm#can (accessed March 25, 2005).

11. Sales include both wholesale and retail merchants.

12. Rosara Lucy Passero, "Ethnicity in the Men's Ready-Made Clothing Industry, 1880–1950: The Italian Experience in Philadelphia" (Ph.D. thesis, University of Pennsylvania, 1978); Salvatore Amico, *Gli Italiani e L'Internazionale dei Sarti da Donna: Raccolta di Storie e Memorie Contemporanee* (Mamaronek, NY: n.p., 1944).

13. Concentration within a given industry, of course, does not indicate concentration within particular workplaces. Industries without ethnic niches might nonetheless have been characterized by ethnically endogamous individual workplaces.

14. A. S., interviewed by author, Philadelphia, PA, September 19, 2007.

15. C. C., interviewed by author, Philadelphia, PA, September 19, 2007.

16. On generational divide within the twentieth-century Italian American population, see Simone Cinotto, "Leonard Covello, the Covello Papers, and the History of Eating Habits among Italian Immigrants in New York," *Journal of American History* 91, no. 2 (September 2004): 497–521; Robert A. Orsi, *Thank You, St. Jude: Women's Devotion to the Patron Saint of Hopeless Causes* (New Haven, CT: Yale University Press, 1996), 9–14, idem, "Conflicts in the Domus," in *The Madonna of 115th Street: Faith and Community in Italian Harlem, 1880–1950* (1985; New Haven, CT: Yale University Press, 2002), chap. 5.

17. C. S., interviewed by author, Philadelphia, PA, September 17, 2007.

18. C. C., interviewed by author, Philadelphia, PA, September 19, 2007.

19. Joel Perlmann and Mary C. Waters, "Intermarriage Then and Now: Race, Generation, and the Changing Meaning of Marriage," in *Not Just Black and White: Historical and Contemporary Perspectives on Immigration, Race, and Ethnicity in the United States,* ed. Nancy Foner and George M. Fredrickson (New York: Russell Sage Foundation, 2004), 272–273; Richard D. Alba, *Italian Americans: Into the Twilight of Ethnicity* (Englewood Cliffs, NJ: Prentice-Hall, 1985).

20. On new immigrant niches in the postwar era, see Waldinger, *Still the Promised City?*

21. In 1951 and 1961, I use aggregate data that was provided on special order by Statistics Canada to Franca Iacovetta; see *Such Hardworking People,* appendix, tables 11–14. Iacovetta's data include a detailed industry variable, facilitating the identification of niches.

22. Iacovetta, *Such Hardworking People,* 47.

23. Iacovetta, *Such Hardworking People,* 24–49.

24. Mirella Borsoi, interviewed by Doreen Rumack, Toronto, ON, March 24, 1988, Folklife Fonds, Archives of Ontario, Toronto.

25. Classified, January 19 1960, *Corriere Canadese* (Toronto).

26. *Ragazzo, ragazza* translates as "young man," "young woman," respectively. Classified, February 07 1961, *Corriere Canadese* (Toronto).

27. Classified, February 07 1961, *Corriere Canadese* (Toronto)

28. "Italian women" or "mothers of families." Classified, January 5 1960 and January 19 1960, *Corriere Canadese* (Toronto).

29. Giovanni De Marsico, interview by Robin Healey, Toronto, ON, March 21, 1979, MHSO.

30. Iacovetta, *Such Hardworking People,* 94–99; Franc Sturino, "The Role of Women in Italian Immigration to the New World," in *Looking into My Sister's Eyes: An Exploration in Women's History,* ed. Jean Burnet, 21–32 (Toronto: Multicultural History Society of Ontario, 1986).

31. Paul N., interview by author, Brampton, ON, November 22, 2007.

32. Francesca S., interview by author, Scaborough, ON, November 22, 2007. This account is corroborated by the observations of historians in other contexts. See Virginia Yans-McLaughlin, "Like Fingers of the Hand: Patterns of Work and Family Organization," in *Family and Community: Italian Immigrants in Buffalo, 1880–1930* (Ithaca: Cornell University Press, 1977), chap. 7

33. Paul N., interview by author, Brampton, ON, November 22, 2007.

34. Paul N., interview by author, Brampton, ON, November 22, 2007.

35. See chapter 1.

36. U.S. Bureau of the Census, *1980 Census of Population and Housing, Census Tracts, Philadelphia, PA-NJ Standard Metropolitan Statistical Area,* figure P-8: "Ancestry of Persons" (Washington, DC: Bureau of the Census, 1983). This figure, like all analysis of Philadelphia tract data to follow, uses the two census tracts that make up Annunciation as well as the tract in the eastern, Italian section of St. Thomas Parish. See chapter 1 for a discussion of the contours of Italian South Philadelphia.

37. Unfortunately, metropolitan microdata and tract level aggregate data match poorly for this analysis. The metropolitan figure is not available in 1960 because of the absence of geographic identifiers for cities in the public use sample of that year. Digitized tract-level aggregate data, meanwhile, did not provide a breakdown of the labor force by industry in 1950. Hence, these reflections use the microdata from 1950 and the tract-level data from 1960. The metropolitan figure cited here, from 1950, is likely somewhat higher than it would have been in 1960.

38. For the causes and ramifications of this process, see Carolyn Adams

et al., *Philadelphia: Neighborhoods, Division, and Conflict in a Postindustrial City* (Philadelphia: Temple University Press, 1991); Philip Scranton, "Large Firms and Industrial Restructuring: The Philadelphia Region, 1900–1980," *Pennsylvania Magazine of History and Biography* 116, no. 4 (1992): 419–465.

39. Department of Public Property, *Plan for South Philadelphia Subway Extension* (City of Philadelphia, 1959).

40. Philadelphia City Planning Commission, *Traffic Feasibility Study of a Chestnut Street Mall in Center City Philadelphia* (Philadelphia: 1966), 11–12, 15.

41. Defined here, as in chapter 4, as the area bounded in the south by Oregon Avenue, the north by South Street, the east by Sixth Street, and the west by Twenty-third Street

42. The total figures here vary slightly from those reported in the sections that work with tract-level aggregate data because the microdata refer to the entire Census Metropolitan area rather than the city of Philadelphia alone. These figures are consistent with the microdata provided for metropolitan areas in the section on niche industries above.

43. In 1990, Italian ethnicity and gender interact in a fashion that is statistically significant but socially negligible. In that census, the mean difference between male and female travel times was slightly larger for Italians than for non-Italians, but the difference in this regard was only one minute.

44. Figures complied from Metropolitan Toronto Planning Board, *Metropolitan Toronto Key Facts* (Toronto: Metropolitan Toronto Planning Board, 1970), table 28; Christopher R. Bryant and Philip M. Coppack, "The City's Countryside," in *Canadian Cities in Transition,* ed. Trudi Bunting and Pierre Filion (Toronto: Oxford University Press, 1991), 232; Pierre Filion and Tod Rutherford, "Employment Transitions in the City," in *Canadian Cities in Transition,* ed. Trudi Bunting and Pierre Filion, 2nd ed. (Toronto: Oxford University Press, 2000), 365; Richard Shearmur and William J. Coffey, "A Tale of Four Cities: Intrametropolitan Employment Distribution in Toronto, Montreal, Vancouver, and Ottawa-Hull, 1981–1996," *Environment and Planning A* 34 (2002): 575–598.

45. The following analysis derives from data files not used in the rest of the chapter: University of Toronto Data Library Service, Census of Canada, 1971, Basic Summary Tabulations, "Place (census tract) of residence in 1971 by place (census tract) of work in 1971," http://www.chass.utoronto.ca.myaccess.library.utoronto.ca/datalib/cc71/cc71bsts.htm#powctlf (accessed March 25, 2005); University of Toronto Data Library Service, Census of Canada, 1981, Basic Summary Tabulations, "Employed labour force by sex (3) for place of residence (each CT and remainder) (location AA) by place of work (each CT, at home, no usual place of work, outside C(M)A, outside Canada and CT not stated (location BB), 1981," http://www.chass.utoronto.ca.myaccess.library.utoronto.ca/datalib/cc81/cc81bsts_ct.htm (accessed March 25, 2005).

46. In 1981 the place of work variables are missing some tracts on the periphery of the GTA.

47. The distance from the far western edge of the parish to the far eastern edge of the employment hub was just over 2.5 miles.

48. The tract-level data on place of work in Toronto divide the workforce by gender; as noted above this is not the case in Philadelphia.

49. Susan Hanson and Geraldine Pratt, *Gender, Work, and Space* (London: Routledge, 1995); Susan Hanson and Ibipo Johnston, "Gender Differences in Work-Trip Length: Explanations and Implications," *Urban Geography* 6, no. 3 (1985): 193–219; Janice Fanning Madden, "Why Women Work Closer to Home," *Urban Studies* 18 (1981): 181–194. For a contrasting perspective, see Kim England, "Suburban Pink Collar Ghettos: The Spatial Entrapment of Women?" *Annals of Association of American Geographers* 83, no. 2 (June 1993): 225–242. On Toronto's Little Italy see, Iacovetta, *Such Hardworking People,* 94, 97–99.

50. In Canada, see Bruno Ramirez and Michael Del Balso, *The Italians of Montreal: From Sojourning to Settlement, 1900–1921* (Montreal: Associazione di Cultura Populare Italo-Quebecchese, 1980); John Zucchi, *The Italian Immigrants of the St. John's Ward, 1877–1915: Patterns of Settlement and Neighborhood Formation* (Toronto: Multicultural History Society of Ontario, 1980); Orest T. Martynowych, *Ukrainians in Canada: The Formative Period, 1891–1924* (Edmonton: Canadian Institute of Ukrainian Studies, 1991), 138–142; Carmela Patrias, *Patriots and Proletarians: Politicizing Hungarian Immigrants in Interwar Canada* (Montreal: McGill–Queen's University Press, 1994), 75–94; Lillian Petroff, *Sojourners and Settlers: The Macedonian Community of Toronto to 1940* (Toronto: University of Toronto Press, 1995). For a survey of the literature in Canada, see Franca Iacovetta, *The Writing of English Canadian Immigrant History* (Ottawa: Canadian Historical Association, 1997), 12–16. In the United States, see John Bodnar, *The Transplanted: A History of Immigrants in Urban America* (Bloomington: Indiana University Press, 1985); Ewa Morawska, *For Bread with Butter: Life Worlds of East Central Europeans in Johnstown, Pennsylvania, 1890–1940* (Cambridge: Cambridge University Press, 1985); Dino Cinel, *From Italy to San Francisco: The Immigrant Experience* (Stanford, CA: Stanford University Press, 1982); Yans-McLaughlin, *Family and Community;* and Humbert Nelli, *Italians in Chicago, 1880–1930: A Study in Ethnic Mobility* (New York: Oxford University Press, 1970). For broad considerations of the role of work and niches in the making of ethnicity, see Kathleen Neils Conzen et al., "The Invention of Ethnicity: A Perspective from the USA," *Journal of American Ethnic History* 12, no. 1 (1992): 3–41; Michael Hechter, "Group Formation and the Cultural Division of Labor," *American Journal of Sociology* 84 (1978): 293–319; William Yancey, Eugene Ericksen, and Richard Juliani, "Emergent Ethnicity: A Review and Reformulation," *American Sociological Review* 41 (June 1976): 391–403.

51. For an exploration of the split between home and work in American cities, see Ira Katzneslson, *City Trenches: Urban Politics and the Patterning of Class in the United States* (New York: Pantheon Books, 1981).

CONCLUSION

1. Stephen Muzzatti, "Imagining Woodbridge from Below: An Autoethnography of Class and 'Italian-ness' in 1980s Toronto," paper presented at

Project Vaughan: Toward an Understanding of the Italian Canadian Experience Beyond "Little Italy," Toronto 2007.

2. John Bodnar, *The Transplanted: A History of Immigrants in Urban America* (Bloomington: Indiana University Press, 1985). For a compelling and theoretically explicit case study, see Ewa Morawska, *Insecure Prosperity: Small-Town Jews in Industrial America, 1890–1940* (Princeton, NJ: Princeton University Press, 1996).

3. Samuel L. Baily, *Immigrants in the Lands of Promise: Italians in Buenos Aires and New York City, 1870–1914* (Ithaca: Cornell University Press, 1999), 12. Other comparative studies include Donna Gabaccia, *Italy's Many Diasporas* (Seattle: University of Washington Press, 2000); Walter Nugent, *Crossings: The Great Transatlantic Migrations, 1870–1914* (Bloomington: University of Indiana Press, 1992); Donna Gabaccia, *Militants and Migrants: Rural Sicilians Become American Workers* (New Brunswick, NJ: Rutgers University Press, 1988); Dominique Schnapper, "Jewish Minorities and the State in the United States, France, and Argentina," in *Center: Ideas and Institutions,* ed. Liah Greenfeld and Michael Mertin, 186–209 (Chicago: University of Chicago Press, 1988); Herbert S. Klein, "The Integration of Italian Immigrants in the United States and Argentina: A Comparative Analysis," *American Historical Review* 88 (April 1983): 306–346; Caroline B. Brettell, "Is the Ethnic Community Inevitable? A Comparison of the Settlement Patterns of Portuguese Immigrants in Toronto and Paris," *Journal of Ethnic Studies* 9 (Fall 1981): 1–17; Nancy Foner, "West Indians in New York City and London: A Comparative Analysis," *International Migration Review* 13 (Summer 1979): 284–297; John Walker Briggs, *An Italian Passage: Immigrants to Three American Cities, 1890–1930* (New Haven, CT: Yale University Press, 1978); Andrew S. Reutlinger, "Reflections on the Anglo-American Jewish Experience: Immigrants, Workers, and Entrepreneurs in New York and London, 1870–1914," *American Jewish Historical Quarterly* 66 (June 1977): 473–484.

4. Kathleen Neils Conzen, David A. Gerber, Ewa Morawska, George E. Pozzetta, and Rudolph J. Vecoli, "The Invention of Ethnicity: A Perspective from the United States," *Journal of American Ethnic History* 12, no. 1 (1992): 4–5. See also Nancy Foner and George M. Fredrickson, "Immigration, Race, and Ethnicity in the United States: Social Constructions and Social Relations in Historical and Comparative Perspective," in *Not Just Black and White: Historical and Contemporary Perspectives on Immigration, Race, and Ethnicity in the United States,* ed. Nancy Foner and George M. Fredrickson (New York: Russell Sage Foundation, 2004).

5. Rogers Brubaker, "Ethnicity without Groups," *European Journal of Sociology* 43, no. 2 (August 2002): 168.

6. Brubaker, "Ethnicity without Groups," 168. Unlike Brubaker, who emphasizes the political and cognitive events that constitute ethnicity, I focus on the social organizational events that bring members of a group together, constituting ethnicity in everyday life.

7. Richard Alba, *Ethnic Identity: The Transformation of White America* (New Haven, CT: Yale University Press, 1990); Jeffrey G. Reitz and Raymond

Breton, *The Illusion of Difference: Realities of Ethnicity in Canada and the United States* (Ottawa: C. D. Howe Institute, 1994).

8. Thomas A. Guglielmo, *White on Arrival: Italians, Race, Color, and Power in Chicago, 1890–1945* (New York: Oxford University Press, 2003). See also Russell Kazal, *Becoming Old Stock: The Paradox of German-American Identity* (Princeton: Princeton University Press, 2004); Jennifer Guglielmo and Salvatore Salerno, eds., *Are Italians White? How Race Is Made in America* (New York: Routledge, 2003); Matthew Frye Jacobson, *Whiteness of a Different Color: European Immigrants and the Alchemy of Race* (Cambridge, MA: Harvard University Press, 1998); John T. McGreevy, *Parish Boundaries: The Catholic Encounter with Race in the Twentieth-Century Urban North* (Chicago: University of Chicago Press, 1996), chap. 2; Noel Ignatiev, *How the Irish Became White* (New York: Routeledge, 1995); David R. Roediger, *Working Toward Whiteness: How America's Immigrants Became White* (New York: Basic Books, 2005, idem., *The Wages of Whiteness: Race and the Making of the American Working Class* (New York: Verso, 1991). For critical assessments of this literature, see Ewa Morawska, "The Sociology and History of Immigration: Reflections of a Practitioner," in *International Migration Research: Constructions, Omissions, and the Promises of Interdisciplinarity,* ed. Michael Bommes and Ewa Morawska (Hampshire, UK: Ashgate, 2005); Eric Arnesen, "Whiteness and the Historians' Imagination," *International Labor and Working-Class History* 60 (Fall 2001): 3–32; Peter Kolchin, "Whiteness Studies: The New History of Race in America," *Journal of American History* 89, no. 1 (June, 2002): 154–173.

9. My conception of structure and agency draws on Ewa Morawska, "Structuring Migration: The Case of Polish Income-Seeking Travelers to the West," *Theory and Society* 30, no. 1 (2001): 47–80; Mustafa Emirbayer and Ann Mische, "What Is Agency?" *American Journal of Sociology* 103, no. 4 (January 1988): 962–1023; and William H. Sewell Jr., "A Theory of Structure: Duality, Agency, and Transformation," *American Journal of Sociology* 98, no. 1 (July 1992): 1–29.

10. Arnold R. Hirsch, "E Pluribus Duo? Thoughts on 'Whiteness' and Chicago's 'New' Immigration as a Transient Third Tier," *Journal of American Ethnic History* 23, no. 4 (Summer 2004): 7–44; James R. Grossman, *Land of Hope: Chicago, Black Southerners and the Great Migration* (Chicago: University of Chicago Press, 1989).

11. Ninette Kelly and Michael Trebilcock, *The Making of the Mosaic: A History of Canadian Immigration Policy* (Toronto: University of Toronto Press, 1998), chaps. 8–9; Roger Daniels, "The Cold War and Immigration," and "Lyndon Johnson and the End of the Quota System," in *Guarding the Golden Door: American Immigration Policy and Immigration since 1882* (New York: Hill and Wang, 2004), chaps. 6–7; George J. Borjas, "Immigration Policy, National Origin, and Immigration Skills: A Comparison of Canada and the United States," in *Small Differences That Matter: Labor Markets and Income Maintenance in Canada and the United States,* ed. David Card and Richard B. Freeman, 21–43 (Chicago: University of Chicago Press, 1993).

12. Steven C. High, *Industrial Sunset: The Making of North America's Rust Belt, 1969–1984* (Toronto: University of Toronto Press, 2003).

13. Nicholas Demaria Harney, *Eh Paesan! Being Italian in Toronto* (Toronto: University of Toronto Press, 1998), chap. 4.

14. Even scholars who seek to draw a sharp divide between urban experience in Canada and the United States acknowledge the tremendous diversity of cities in each national context. See: Michael A. Goldberg and John Mercer, *The Myth of the North American City: Continentalism Challenged* (Vancouver: University of British Columbia Press, 1986).

15. Jeremy Boissevain, *The Italians of Montreal: Social Adjustment in a Plural Society* (Ottawa: Studies of the Royal Commission on Bilingualism and Biculturalism, 1970), 33. Boissevain's suggestion dovetails with other grounds for thinking the social geography of Montreal would differ from that in Toronto. Ethnic groups and the poor were significantly more segregated in Montreal than in Toronto. For comparative data on ethnic segregation in each context, see my Web site: http://citystats.uvic.ca/.

16. Nicholas Demaria Harney, "Ethnicity, Social Organization, and Urban Space: A Comparison of Italians in Toronto and Montreal," in *Urban Enigmas: Montreal, Toronto, and the Problem of Comparing Cities,* ed. Johanne Sloan (Montreal: McGill–Queen's University Press, 2007), 204.

17. Kevin Michael Schultz, "The Decline of the Melting Pot: Catholics, Jews, and Pluralism in Postwar America" (Ph.D. thesis, University of California, Berkeley, 2005), 242; Matthew Frye Jacobson, *Whiteness of a Different Color: European Immigrants and the Alchemy of Race* (Cambridge, MA: Harvard University Press, 1999), 91–135; Roediger, *Working Toward Whiteness,* 133–234; Lizabeth Cohen, *Making a New Deal: Industrial Workers in Chicago, 1919–1939* (New York: Cambridge University Press, 1990); Gary Gerstle, *Working-Class Americanism: The Politics of Labor in a Textile City, 1914–1960* (Cambridge: Cambridge University Press, 1989); Steve Fraser and Gary Gerstle, *The Rise and Fall of the New Deal Order, 1930–1980* (Princeton: Princeton University Press, 1989); Lizabeth Cohen, *A Consumers' Republic: The Politics of Mass Consumption in Postwar America* (New York: Alfred Knopf, 2003).

18. Schultz, "The Decline of the Melting Pot"; Thomas J. Sugrue and John D. Skrentny, "The White Ethnic Strategy," in *Rightward Bound: Making American Conservative in the 1970s,* ed. Bruce J. Schulman and Julian E. Zelizer (Cambridge, MA: Harvard University Press, 2008).

19. See introduction, note 10.

20. Stefano Luconi, *From Paesani to White Ethnics: The Italian Experience in Philadelphia* (Albany: State University of New York Press, 2001), 126. Luconi offers a fine example of why analysis of the emergence of whiteness need not conflict with my claims about the ongoing social organizational importance of ethnicity. While Luconi is not concerned with neighborhood per se, his analysis suggests continuities between white ethnicity and more particularized Italian ethnic identity and political claims.

21. Roediger, *Working Toward Whiteness,* 166. Matthew Frey Jacobson similarly observes that "ethnic particularism has been among the idioms of

white backlash" ("Hyphen Nation: Ethnicity in American Intellectual and Political Life" in *A Companion to Post-1945 America,* ed. Jean-Christophe Agnew and Roy Rosenzweig [Malden, MA: Blackwell, 2006], 188).

22. Gerald H. Gamm, *Urban Exodus: Why the Jews Left Boston and the Catholics Stayed* (Cambridge, MA: Harvard University Press, 1999); John T. McGreevy, *Parish Boundaries: The Catholic Encounter with Race in the Twentieth-Century Urban North* (Chicago: University of Chicago Press, 1996), 19–28.

23. Gamm, *Urban Exodus,* 17.

24. Hirsch, "Friends, Neighbors, and Rioters," in *Making the Second Ghetto: Race and Housing in Chicago, 1940–1960* (1983; Chicago: University of Chicago Press, 1998), *Making the Second Ghetto,* 68–99; quotation from 99.

25. Sugrue, *Origins of the Urban Crisis,* 235–258; Ronald P. Formisano, *Boston against Bussing: Race, Class, and Ethnicity in the 1960s and 1970s* (Chapel Hill: University of North Carolina Press, 1991).

26. See also Jonathan Rieder, *Canarsie: The Jews and Italians of Brooklyn against Liberalism* (Cambridge, MA: Harvard University Press), 32–43; McGreevy, *Parish Boundaries,* 91–110; Eileen M. McMahon, *What Parish Are You From? A Chicago Irish Community and Race Relations* (Lexington: University Press of Kentucky, 1995).

27. Robert Anthony Orsi, *The Madonna of 115th Street: Faith and Community in Italian Harlem, 1880–1950,* 2nd ed. (New Haven, CT: Yale University Press, 2002), xx.

28. For a fascinating example of the malleable spatial structure of religious practice in another context, see Robert Orsi, "The Centre Out There, In Here, and Everywhere Else: The Nature of Pilgrimage to the Shrine of St. Jude, 1929–1965," *Journal of Social History* 25 (1991): 213–232. Elsewhere, I have expressed qualms with the notion that institutional rules predetermined the shape of Jewish religious life: Jordan Stanger-Ross, "Neither Fight nor Flight: Philadelphia Synagogues in an Era of Suburbanization," *Journal of Urban History* 32, no. 6 (2006): 791–812.

29. Stephen A. Speisman, *The Jews of Toronto: A History to 1937* (Toronto: McClelland Stewart, 1979), 81–95; Bruno Ramirez and Michael Del Balso, *The Italians of Montreal: From Sojourning to Settlement, 1900–1921* (Montreal: Associazione di Cultura Populare Italo-Quebecchese, 1980); John Zucchi, *The Italian Immigrants of the St. John's Ward, 1877–1915: Patterns of Settlement and Neighborhood Formation* (Toronto: Multicultural History Society of Ontario, 1980); Robert F. Harney and J. Vincenza Scarpaci, eds., *Little Italies in North America* (Toronto: Multicultural History Society of Ontario, 1981); Robert F. Harney, ed., *Gathering Place: Peoples and Neighborhoods of Toronto, 1834–1945* (Toronto: Multicultural History Society of Ontario, 1985); Patricia Roy, *A White Man's Province: British Columbia Politicians and Chinese and Japanese Immigrants, 1858–1914* (Vancouver: University of British Columbia Press, 1989); Kay Anderson, *Vancouver's Chinatown: Racial Discourse in Canada, 1875–1980* (Montreal: McGill–Queen's University Press, 1991); Orest T. Martynowych, *Ukrainians in Canada: The*

Formative Period, 1891–1924 (Edmonton: Canadian Institute of Ukrainian Studies, 1991), 138–142; Gerald Tulchinsky, *Taking Root: The Origins of the Canadian Jewish Community* (Toronto: Lester Publishing, 1992); Franca Iacovetta, *The Writing of English Canadian Immigrant History* (Ottawa: Canadian Historical Association, 1997), 12–16; Carmela Patrias, "Hungarian Communities in Canada," in *Patriots and Proletarians: Politicizing Hungarian Immigrants in Interwar Canada* (Montreal: McGill–Queen's University Press, 1994), 75–94; Lillian Petroff, *Sojourners and Settlers: The Macedonian Community of Toronto to 1940* (Toronto: University of Toronto Press, 1995); Megan J. Davies, "Night Soil, Cesspools, and Smelly Hogs on the Streets: Sanitation, Race, and Governance in Early British Columbia," *Histoire Sociale/Social History* 38, no. 75 (May 2005): 1–36.

30. John E. Zucchi "Italian Hometown Settlements and the Development of an Italian Community in Toronto, 1875–1935," in Harney, *Gathering Place,* 123.

31. Sturino, *Forging the Chain;* Zucchi, *Italians in Toronto;* Robert F. Harney, "Ethnicity and Neighborhoods," in Harney, *Gathering Place,* 1–24; Zucchi, "Italian Hometown Settlements."

32. Zucchi, "Italian Hometown Settlements," 143–144.

33. Franca Iacovetta, *Such Hardworking People: Italian Immigrants in Postwar Toronto* (Montreal: McGill–Queen's University Press, 1992), 56.

34. Richard H. Thompson, *Toronto's Chinatown: The Changing Social Organization of an Ethnic Community* (New York: AMS Press, 1989), 186–198; Anderson, *Vancouver's Chinatown,* chap. 6; Peter S. Li, *The Chinese in Canada* (Toronto: Oxford University Press, 1998), 112–116; Murdie and Teixiera, "Towards a Comfortable Neighbourhood and Appropriate Housing," in *The World in a City,* ed. Paul Anisef and C. Michael Laniphier (Toronto: University of Toronto Press, 2003), 154; David Lai, *Chinatowns: Towns within Cities in Canada.* (Vancouver: University of British Columbia Press, 1988).

35. Peter D. Chimbos, *The Canadian Odyssey: The Greek Experience in Canada* (Toronto: McClelland and Stewart, 1980), 72–88; Brettell, "Is the Ethnic Community Inevitable?; Carlos Teixeira, "On the Move: Portuguese in Toronto," in *The Portuguese in Canada: From Sea to the City,* ed. Carlos Teixeira and Victor M. P. Da Rosa, 207–220 (Toronto: University of Toronto Press, 2000); Murdie and Teixiera, "Towards a Comfortable Neighbourhood and Appropriate Housing," 158–161.

36. Masumi Izumi, "Reconsidering Ethnic Culture and Community: A Case Study on Japanese Canadian Taiko Drumming," *Journal of Asian American Studies* 4, no. 1 (2001): 40–41.

37. Daniel Hiebert, "Immigration and the Changing Social Geography of Greater Vancouver," BC Studies 121 (Spring 1999): 79.

38. Clifford Jansen and Lawrence Lam, "Immigrants in the Greater Toronto Area: A Sociodemographic Overview," in Anisef and Laniphier, *World in a City,* 67.

Index

Page numbers followed by "f" indicate figures; page numbers followed by "m" indicate maps.

Series titles, continued from frontmatter

City of American Dreams: A History of Home Ownership and Housing Reform in Chicago, 1871–1919 *by Margaret Garb*

Chicagoland: City and Suburbs in the Railroad Age *by Ann Durkin Keating*

The Elusive Ideal: Equal Educational Opportunity and the Federal Role in Boston's Public Schools, 1950–1985 *by Adam R. Nelson*

Block by Block: Neighborhoods and Public Policy on Chicago's West Side *by Amanda I. Seligman*

Downtown America: A History of the Place and the People Who Made It *by Alison Isenberg*

Places of Their Own: African American Suburbanization in the Twentieth Century *by Andrew Wiese*

Building the South Side: Urban Space and Civic Culture in Chicago, 1890–1919 *by Robin F. Bachin*

In the Shadow of Slavery: African Americans in New York City, 1626–1863 *by Leslie M. Harris*

My Blue Heaven: Life and Politics in the Working-Class Suburbs of Los Angeles, 1920–1965 *by Becky M. Nicolaides*

Brownsville, Brooklyn: Blacks, Jews, and the Changing Face of the Ghetto *by Wendell Pritchett*

The Creative Destruction of Manhattan, 1900–1940 *by Max Page*

Streets, Railroads, and the Great Strike of 1877 *by David O. Stowell*

Faces along the Bar: Lore and Order in the Workingman's Saloon, 1870–1920 *by Madelon Powers*

Making the Second Ghetto: Race and Housing in Chicago, 1940–1960 *by Arnold R. Hirsch*

Smoldering City: Chicagoans and the Great Fire, 1871–1874 *by Karen Sawislak*

Modern Housing for America: Policy Struggles in the New Deal Era *by Gail Radford*

Parish Boundaries: The Catholic Encounter with Race in the Twentieth-Century Urban North *by John T. McGreevy*